John Good

**Measuring Made Easy**

Or, the description and use of Coggeshall's sliding rule.

John Good

**Measuring Made Easy**
*Or, the description and use of Coggeshall's sliding rule.*

ISBN/EAN: 9783337175351

Printed in Europe, USA, Canada, Australia, Japan

Cover: Foto ©berggeist007 / pixelio.de

More available books at **www.hansebooks.com**

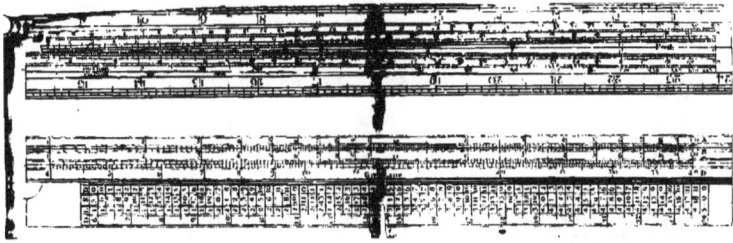

# MEASURING made EASY,

### Or the Defcription and Ufe of

# Coggefhall's Sliding Rule:

#### Containing Inftruction for

## Meafuring all Manner of TIMBER,

Both by the Common Way, and the true Way:

With Directions for taking the Dimenfions of Trees, and the Allowance for Bark, &c.

Performed both by the Rule, and by Arithmetick;

By which may be meafured

## All MANNER of SUPERFICIES,

As Board, Glafs, Plaiftering, Painting, Wainfcotting, Tyling, Paving, Land, &c, Both by the Rule and Arithmetic.

#### To which is now added,

The Defcription of *Scamozzi*'s Lines, with the Ufe in finding the Length and Angles of Rafters, Hips, Collar-Beams, &c.

---

*By* J. G o o D, Teacher of the Mathematics.

---

Carefully Corrected, and much Enlarg'd . By J. A T K I N S O N.

---

*L O N D O N:*

Printed for MOUNT & PAGE, *Tower-hill,* 1786.

# TO THE
# READER.

I Here present you with some Fruits of my spare Hours, wherein I have endeavoured to render the Use of the Sliding-Rule more Plain and Easy than ever it appeared before.

I have here delivered in a few Rules, but yet in a Plain and Easy Manner, Problems for the measuring all Manner of Regular Figures, and Plain Superficies, such as Board, Glass, Painting, Plaistering, Paving, Tyling, Joyners and Masons *Work*, &c. and that both by the Sliding-Rule, and by Arithmetick.

In the next Place, you have Directions for the measuring all Manner of Timber or Stone, by the Sliding-Rule, and Arithmetically, by a new Method, whereby the true Content of Round Timber or Trees, may be found.

I remain, a *Well-Wisher* to the Mathematicks,

## JOHN GOOD.

# CHAP. I.

## *The Description and Use of* Coggeſhall's RULE.

THIS *Rule* is framed three Ways; *Sliding* by one another, like *Glazier's Rule*; *Sliding* on one Side of a *Two-Foot Joint Rule*; and one Part *Sliding* on the other, in a Foot long, the back Part being flat, on which are fundry *Lines* or *Scales*.

Upon the forefaid *Sliding Side* of the Rule, are four Lines of *Numbers*, three are *Double Lines*, and one a fingle *Line of Numbers*; marked in the *Figure* annexed, with A, B, C, and D.

The three marked A, B, and C, are called *Double Lines of Numbers*; being figured 1, 2, 3, 4, 5, 6, 7, 8, 9. Then, 1, 2, 3, 4, 5, 6, 7, 8, 9, and 10 at the End.

That marked D, is the *Single Line of Numbers*, and figured thus, 4, 5, 6, 7, 8, 9, 10, 20, 30, and 40 at the End, even with and under 10 in the *Double Line* next to it, this is called the *Girt-line*, and is fo marked in the *Figure*.

The

The Figures on the Three *Double Lines of Numbers* may be encreafed or decreafed at pleafure; thus, 1 at the Beginning may be called 10, 100, or 1000, then 2 is 20, 200, or 2000, fo that when 1 at the Beginning is 10, then 1 in the middle is 100, and then 10 at the End is 1000; but if 1 at the Beginning is counted for 1, then 1 in the Middle is 10, and 10 at the End is 100.

And as the Figures are altered, fo muft the Strokes or Divifions between them be altered in their Value, according to the Number of the Parts they are divided into : As thus, the Diftance from 1 to 2 is divided into 10 Parts, and each Tenth is divided into 5 Parts; and that from 2 to 3 is divided into 10 Parts; and each Tenth into 2 Parts, and fo on from 3 to 5 ; then the Diftance from 5 to 6, is divided into 10 Parts only ; and fo on to 1 in the Middle of the *Rule*, or the Firft Part of the *Double Line of Numbers*. The Second Part, or *Radius*, is divided like the *Firft Radius*.

The *Girt* Line marked D, is divided from 4 to 5 into 10 Parts, and each Tenth into 2 Parts, and fo on from 5 to 10, then from 10 to 20, it's divided into 10 Parts, and each Tenth is divided into 4 Parts ; and fo on all the way from 20 to 40 at

the

the End, which is right againft 10 at the End of the *Double Line of Numbers.*

The Lines on the Back-fide of this *Rule* that flide on the one Side, are thefe, a Line of *Inch Meafure* from 1 to 12, each Inch divided into Halfs, Quarters, and Half-Quarters; another Line of *Inch Meafure* from 1 to 12, each Inch divided into 10 equal Parts; and a *Line of Foot Meafure*, being one Foot divided into 100 equal Parts, and Figured 10, 20, 30, 40, 50, 60, 70, 80, 90 and 100 even with 12 on Inch Meafure.

And the Back-fide of the *Sliding Piece* is divided into Inches, Halves, Quarters, and Half Quarters, and figured from 12 to 24; fo that it may be flid out to 2 Feet to meafure the Length of a Tree, or any Thing elfe you have occafion to meafure.

*Note,* The *Line of Foot-meafure* turns Inches, Halves, Quarters, &c. into Decimal Parts of a Foot, which is moft ready for meafuring by this *Rule*, as alfo by Arithmetick; which is fully explained in the Table following.

*A* T A B L. E *Reducing Inches, Halves, Quarters, and Half-Quarters, into Decimal Parts of a Foot; Or, reducing Inch-Measure into Foot Measure.*

| Inches | Decimal | Parts of an Inch. | | | | | | |
|---|---|---|---|---|---|---|---|---|
| | | ⅛ | ¼ | ⅜ | ½ | | ¾ | ⅞ |
| 0 | .000 | .010 | .021 | .031 | .042 | .052 | .062 | .073 |
| 1 | .083 | .094 | .104 | .115 | .125 | .135 | .146 | .156 |
| 2 | .167 | .177 | .188 | .198 | .208 | .219 | .229 | .24 |
| 3 | .25 | .26 | .271 | .281 | .292 | .302 | .313 | .323 |
| 4 | .333 | .344 | .35 | .365 | .275 | .385 | .396 | .406 |
| 5 | .417 | .427 | .438 | .448 | .458 | .469 | .479 | .49 |
| 6 | .5 | .51 | .521 | .531 | .542 | .552 | .569 | .573 |
| 7 | .583 | .594 | .604 | .615 | .625 | 6.35 | .646 | .656 |
| 8 | .667 | .677 | .688 | .698 | .708 | .719 | .729 | .74 |
| 9 | 75 | .76 | .771 | .781 | .792 | .802 | .813 | .823 |
| 10 | .833 | .844 | .854 | .865 | .875 | .885 | .89 | .906 |
| 11 | .917 | .927 | .937 | .948 | .958 | .969 | 979 | .99 |

By this Table ⅛ or Half Quarter of an Inch in *Foot Measure* in Decimals, is 01, or ₁₀₀ parts of a Foot; ¼ of an Inch is .021, or shorter .02, that is ₁₀₀ parts of a Foot; and so on in the upper Line of the Table. And in the next Line under it, 1 Inch is

A 4                                    .083

.083 or $\frac{8}{100}$ of a Foot; Inch $1\frac{1}{8}$ is .094 or $\frac{9}{100}$; Inch $1\frac{1}{4}$ is, .104 or $\frac{10}{100}$; Inch $1\frac{3}{8}$ is .115, or $\frac{15}{100}$ of a Foot and so on; And in the First Column under Decimals 1 Inch is .083; 2 Inches .167; 3 Inches is .25; 4 Inches .333; and so on to 11 Inches is .917 or $\frac{92}{100}$. All these may be found on the *Line of Foot-Measure* on the Back-side of the Rule.

## C H A P. II.

*The Use of the* Double Scale, *shewing how to find the Area of any Plain Superficies; and also* Arithmetically.

### P R O B. I. *To measure a Geometrical Square.*

BY the *Sliding - Rule.* Let there be a Square whose Sides are each three Feet and a half. Set 1 on the Line B, to $3\frac{1}{2}$ on the Line A, then against $3\frac{1}{2}$ on the Line B, is $12\frac{1}{4}$ Feet on the Line A, which is the Content of such a Square.

| 1 | 2 | 3 | a |
|---|---|---|---|
| 4 | 5 | 6 | b |
| 7 | 8 | 9 | f |
| c | d | e | g |

Arith-

Arithmetically.

F.P.

The Side of the Square, Feet 3½ or —— 3.5
Multiplied by itſelf ——— ———— ——— 3.5

175
105.

The Product is ——— ——— ——— Feet 12.25

This Arithmetical Operation is thus performed, multiply the Sides of the Square,
namely 3½, or 3 $\frac{5}{10}$ into itſelf, the Product is
1225, from which cut off 2 Places, becauſe
there are 2 Decimals in both Fractions, and
it will ſtand thus 12 $\frac{25}{100}$; which is 12 Feet
and a Quarter.

*Note*, The Figure itſelf ſhews the Content
in Feet by Inſpection, for if you tell the greater Squares, you may ſee they are 9 in Number;
*a* and *b* makes 10; *c* and *d* makes 11; *e*
and *f* makes 12; and the little Square *g* in
the Corner, makes one Quarter of a Foot,
which is in all 12 Feet.

*More Examples by the Rule.*

Length of the Side   5¼ Feet⎫  27½ Feet
Length of the Side   7½ Feet⎪   56¼ Feet
Length of the Side  10¼ Feet⎪115½ Feet
Length of the Side  13   Feet⎭169  Feet

PROB.

**PROB. II.** *To meafure a* Long-Square *by the* Sliding-Rule ; *and alfo by* Arithmetick.

LET there be a Long Square, whofe longeft Side is 27 Feet and a Half, the fhorteft 16 Feet and a Quarter : fet 1 on the Line B, to 16¼ ·on the Line A, then againft 27½, on the Line B, is 446¼ Feet the Content of the Long Square on the Line A.

### By Decimal Arithmetick.

Multiply the Length by the Breadth, and from the Product cut off as many Places to the Right-hand as there are Decimals in the Length and Breadth ; the Integers remaining to the Left-hand are the Square Feet.

Example. F.Par.

Length 27½ Feet, or————— —— ——— 27.50
Breadth 16¼ Feet, or——— ———— ——— 16.25

$$\begin{array}{r} 13750 \\ 5500 \\ 16500 \\ 2750 \\ \hline 446.8750 \end{array}$$

*But*

*But Contracted thus.*      F. Par.

Breadth Feet 16 ¼ or in Decimals —— 16.25:
Length Feet 27¼ or 27.25, inverted is   5.72

$$\begin{array}{r} 3250 \\ 1137 \\ 81 \\ \hline \end{array}$$

The Product is the Content req. Feet — 446.8

In this *Contracted Way*, Multiply each Figure of the inverted Number (beginning at
the Right-hand) by the Figure over it, and
putting the feveral Multiplications even at
the Right-hand, as above: Thus multiplying by 2, the Product is 3250: Then fay 7
times 5 is 35, leave out 5 and carry 3 in your
Mind, then fay 7 times 2 is 14, and 3 I carried makes 17, fet 7 juft under the 0, and carry the one in your mind; fay 7 times 6 is
42, and 1 I carried makes 43, fet 3 under 5
and carry the 4 in your mind; and fo on to
the End of that Line; Then fay 5 times 2 is
10, reject the 0, and carry one in your mind;
then fay 5 times 6 is 30, and 1 carried makes
31, fet 1 juft under the 7, and carry the 3 in
your mind; faying 5 times 1 is 5, and 3 I
carried makes 8, fetting it under the 3, and
the multiplying is done: Then add them together and it's 446.8; that is Feet 446 $\frac{8}{10}$ or
446 ¾ as before.

*Note.*

*Note,* Where the Units place of the inverted Number ſtands, ſo many places are to be
cut off in the Product; As in this Example,
7 the Units place, when inverted ſtands under
2, the firſt place in Decimals, and therefore
one place muſt be cut off in the Product of the
Right-hand; ſo that the Product 4468 and
be 446.8 or 446 $\frac{8}{10}$; obſerve the like in all that
follows.

*More Examples by the* Sliding-Rule.

Contents

| | | | |
|---|---|---|---|
| Breadth 11 $\frac{1}{2}$ Feet | Length 15 $\frac{1}{4}$ | 174 $\frac{1}{4}$ Feet |
| Breadth 17 $\frac{1}{4}$ Feet | Length 21 $\frac{1}{2}$ | 371 $\frac{3}{4}$ Feet |
| Breadth 22 $\frac{3}{4}$ Feet | Length 19 $\frac{3}{4}$ | 676 $\frac{1}{4}$ Feet |

## PROB. III. *How to meaſure a* Rhombus *by the* Sliding-Rule.

SUppoſe the Side of a *Rhombus* be 8 Feet
6 Inches and a quarter, and the Breadth,
or Line A B, 8 Feet 4 Inches and a half:
That is, the Length is Feet, 8 $\frac{52}{100}$, and Breadth
is Feet $\frac{38}{100}$. Set 1 on the Line B, to 8
Feet $\frac{52}{100}$ on the Line A; then againſt 8 Feet
$\frac{52}{100}$ on the Line B, is 71 Feet $\frac{38}{100}$ parts of a
Foot on the Line A: Now if you would find
the Value of the Decimal, or part of the Foot
look for $\frac{48}{100}$ on your *Rule,* and you will find
againſt it 4 $\frac{3}{4}$ Inches, ſo that the Content of
this *Rhombus* is 71 Feet 4 $\frac{3}{4}$ Inches.

*By*

Multiply the Breadth AB, by the Length of any of the Sides, and from the Product cut off as many Places to the Right-hand as are Decimals in the Length and Breadth, the Integers remaining to the Left-hand are the square Feet required.

*Example.*

|                                                    | *F. P.* |
|----------------------------------------------------|---------|
| The Length Feet 8.06¼ Inches, or —— | 8.52 |
| Breadth A B, Feet 8.04½ Inches, or —— | 8.38 |

$$
\begin{array}{r}
6816 \\
2556 \\
6816 \\
\hline
\end{array}
$$

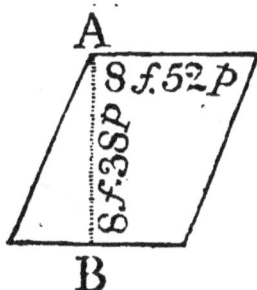

The Content is Feet 71.3976 or 71 Feet 4¾ Inches.

And contracted is thus :

F. P.

Length Feet 8.06 Inches 8¼ is in Decim. 8.52
Breadth A B, Feet 8.04 Inches and ½ is ⎫
8.38, inverted is —— —— —— ⎬ 8.38
⎭

$$
\begin{array}{r}
6816 \\
255 \\
68 \\
\hline
\end{array}
$$

Then *Content* required as before is *Feet* —— 71.39

P R O B·

**PROB. IV.** *How to measure a* Rhomboides
*by the Sliding Rule.*

ADmit the Length of a *Rhomboides* be
17 Feet 3 Inches, or 17 $\frac{15}{100}$, and the
true Breadth 8 Feet 7 Inches, or 8 $\frac{58}{100}$, what
is the Content? Set 1 on the Line B, to 17.25
on the Line A; then against 8.58 on the Line
B, is 148 Feet on the Line A.

*By Decimal Arithmetick.*

Multiply the Length by the Breadth, and
from the Product cut off the Decimals, and
the Remainder is the Square Feet.

|  | *Example.* | F.Par. |
|---|---|---|
| The Length 17 Feet 3 Inches, or —— | 17.25 |
| The Breadth 8 Feet 7 Inches, or —— | 8.58 |

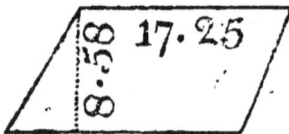

|  |
|---|
| 13800 |
| 8625 |
| 13800 |
| 148.0050 |

The *Content* is Ft. 148.0050
or 148 Feet.

*By Contraction thus,* F.Par.

Length 17 Feet 3 Inches in Decimals —— 17.25
Breadth 8 Ft. 7 Inch. or 8.58, inverted is 85.8

| |
|---|
| 13800 |
| 862 |
| 138 |
| Feet 148.00 |

The Content as before is —— Feet 148.00

PROB.

P. R O B.  V,  *How to meafure a Triangle by the* Sliding-Rule.

*Theorem.* EVery Triangle is half of that Long-Square whofe Length and Breadth is equal to the Perpendicular and Bafe. Therefore from the greateft Angle let fall a Line Perpendicular, to the Side (oppofite to the greateft Angle) called the Bafe; and then to find the Content of any Triangle, take half the Length of the Bafe and the whole Perpendicular; or half the Length of the Perpendicular, and the whole Bafe.

*Example.*

Let the Bafe of a Triangle be 4 Feet, $\frac{14}{100}$ or 4 Feet 1 Inch $\frac{3}{4}$, and the *Perpendicular* 2 Feet $\frac{15}{100}$, or 2 Feet 1 Inch $\frac{3}{4}$, the half of the one is 2 Feet 7 parts, and the half of the other is 1 Foot 7 Parts; fet one on the Line B, to 4 · 5 on the Line A, then againft 1.07, half the *Perpendicular* on the Line B, is 4 Feet and almoft half for the Content. Or if you fet 1 on the Line B, to 1.07 on the Line A; againft 4.1, on the Line B, is 4 and almoft a half on the Line A. Again, another way, if you fet 1 on the Line B, to 4.1 on the Line A; then againft 2.15 on the Line B, is 8 Feet $\frac{9}{10}$ (which is about 11 Inches) on the Line A, the half thereof is 4 Feet 5½ Inches, which is the *Content* of the Triangle.                    *By*

*By Decimal Arithmetick all the three Ways.*

F. P.

**1st Way** { The whole Base ———————4.15
{ The half Perpendicular is — 1.07

2905
415

The Content 4 Ft. 5 ½ Inches or Ft.  4.4405

**2d Way** { The whole Perpendicular — 2.15
{ The half Base ——————— 2.07

1505
430

The Content 4 Feet 5½ Inches, or Feet 4 4505

**3d Way** { The whole Base ———————4.15
{ The whole Perpendicular — 2.15

2075
415
830

The Product is Feet 8.9225

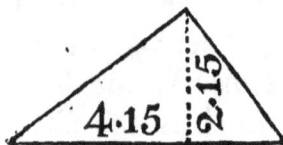

The half is Feet 4.4612
The Content is 4 Feet 5
Inches and ½, as before.

By

*By Contraction thus,*     \F.Par.

The whole Bafe ————— 4.15
Whofe Perpendicular 2.15 inverted is ——51. 2

          830
           41
           21

The Product is ————— 8.92
Half is the Cont. 4 Feet 5½ Inc. as before 4.46

## PROB. VI. *Of the Meafuring a* Trapezium.

A *Trapezium* is a Figure of 4 Sides, but is not a *Square, a Long-fquare,* a *Rhombus,* not a *Rhomboides;* which before it can be meafured muft be divided into Triangles, by a Line from one Corner to his oppofite one, through the *Trapezium;* fo will the two Perpendiculars fall from the other two Angles upon this Diagonal, or Crofs-line; then the Content of the Figures are to be found by the laft *Problem.*

*Example.* Suppofe the Dimenfions of a *Trapezium* were as followeth, the Bafe is 16 Feet 8 Inches, the one Perpendicular 12 Feet 6 Inches, and the other 9 Feet 8 Inches, what is the Content?

       ·B      *Deci-*

*Decimally.*      F. Par.

One Perpendicular 12 Feet 6 Inches or    12.50
The other 9 Feet 8 Inches, or —————   9.68

The Sum is ————————————— 22.18
The half Sum is ——————————— 11.09
Multiply by the whole Base ————— 16.67

        7763
        6654
        6654
       1109

The Content 184 Ft. 10½ Inch. or Ft. 184.8703

By *Contraction* thus :

                F. Par.

Half Sum of the two Perp. 11 F. 1 Inch. or 11.08
Base 16 Feet 8 Inch. or 16.67 inverted is 76.61

         11.08
          665
           67
            8

The Content as before is ————— Feet 184.8

*Note*, If two Sides of a *Trapezium* are parallel, add them together, and halve the Sum, this half Sum multiplied by the nearest Distance betwixt those two Sides, the Product gives the Content. Or if you measure it in the middle betwixt those two Lines that are parallel, it will be the same.

P R O B.

## PROB. VII. *To meafure any Regular Figure.*

OF thefe Regular Figures there are feveral Sorts, as the *Pentagon*, contained under five Sides; the *Hexagon*, contained under fix Sides; the *Heptagon*, contained under feven Sides; and the *Octagon*, contained under eight Sides, &c.

Now to meafure any of thefe Sorts of Figures practically, it is by dividing them into Triangles, which is done by drawing Lines from the Center of the Figure to every Angle, then from the Center to the middle of any one of the Triangle Sides, draw a Line, which Line is the Perpendicular. Having the Perpendicular and Bafe of any of thefe Triangles (by *Prob. 5.*) find the Content of one Triangle, and that multiplied by the Number of Triangles, finds the Content of the Figure.

*Note,* To find the Center of any Regular Figure, of even Number of Sides, draw a Line from one Angle to it's oppofite, the middle of which is the Center; but if your Figure have any odd Number of Sides, as 5, or 7, &c. draw Lines from any two Angles to the middle of their oppofite Sides, their Interfection is the Center,

B 2                                        CHAP.

**C H A P. III.** *The Ufe of the* Girt-line, *and Line of* Numbers, *called the* Double-fcale, *in meafuring of Circles, and their Parts.*

**Prob. i.** LET the Diameter of a Circle be 2 Feet $\frac{25}{100}$, to find the Content, fet 11 on the Girt-line D, to 95 on the Double-line C; then againft 2 Feet $\frac{25}{100}$ on the Girt-line D, is 3 Feet $\frac{28}{100}$ on the Double-fcale of Numbers C, which is the Content.

*Arithmetically.*

*The Rule.* Multiply the Diameter in Inches by itfelf, and then by this fixt Number .7854, and from the Product cut off from the Right-hand 4 Figures, the Remainder to the Left-hand is the Content of the Circle in Inches, which divided by 144, gives the Feet, and the Remainder, if any, divide by 12, gives the Inches. *Example.*

The fixt Number ————————— .7854
The Diameter multiplied ———— 729 Inches

```
                          70686
The Diameter 27 Inches  .  15708
The Diameter 27 Inches     54978
                          ————————  F.In.
        189              144)572.5566( 3. 11
         54                 432
        ————                ————
Diam.multip.729 In.         12)140
                               12
                               ··
                              ————
                              021
                              ————
                              012
                              ————
                               09
```

**The** *Content* **is 3 Feet 11¾ Inches.**          The

*The Rule may be thus,* Multiply the Diameter in *Foot-meafure* by itfelf, the Product is called the *Square of the Diameter,* then multiply by the fixt Number 0.7854 in the *Contracted way,* and this laft Product is the Content required, as follows.                                               F.Par.

The Diameter 2 Feet $\frac{25}{100}$ or ————————2.25
The Diameter 2 Feet $\frac{25}{100}$ inverted is———52.2

```
                                      450
                                       45
                                       11
```

The Square of the Diameter is Feet ——  5.06
The fixt Number 0.7854 inverted is —— 4587
                                      3542

```
                                      405
                                       25
                                        2
```

Content 3 Feet 11¾ Inch. almoft, or Feet 3.974


P R O B. II. *To find the Content of a Semicircle.*

**B**Y having the Diameter by the laft *Problem,* find the Content of the whole Circle, the half thereof is the Content of the *Semicircle.* I think this needs no Example.


P R O B. III.  *To find the Content of a Quarter of a Circle, commonly called a* Quadrant.

*Example-***T**HERE is a Quadrant, whofe Semidiameter is 7 Feet, and
the

the Circuit of the Arch is 11 Feet, what is the
Content? Set 1 on the Double-line B, to 7 the
Semidiameter on the Double-line A, then
against 5 Feet $\frac{5}{100}$ (which is half the Circuit of
the Arch) on B, is $38\frac{1}{2}$ on A, the Content 38
Feet 6 Inches, or 38 $\frac{5}{10}$.

## PROB. IV. *To find the Content of a Sector of a Circle.*

*The Rule.* SET 1 on the Double-line, to
the Semidiameter on the Double-
line, then against half the Circuit of the Arch
on the Double-line, is the Content on the other
Double-line. This needs no *Example.*

## PROB. V. *How to measure the Segment of a Circle.*

A *Segment of a Circle,* is a strait Line drawn
thro' the Circle, but not thro' the Center, divides a Circle into two Parts or Seg-
ments; and the lesser is thus measured Let
the *Sector* be measured, whereof the Segment
is part, then substract the Triangular Part, the
Remainder is the Content of the Segment: But
to the greater Segment, the Content of the Triangle included is to be added.

CHAP.

# C H A P. IV.

*The Ufe of the* Double-Scale; *fhewing how to meafure all manner of Superfices; as Board, Glafs, Painting, Plaiftering, Wainfcotting, Tyling, Paving, &c. by the* Sliding-Rule, *and Arithmetically.*

### Firft. *In meafuring Boards.*

P R O B. I. Let there be a Board whofe Breadth is $27\frac{1}{2}$ Inches, or 27 $\frac{50}{00}$, and the Length 15 Feet $\frac{1}{4}$, or 15 $\frac{25}{00}$, what is the Content by the *Sliding Rule.*

### *Example.*

Set 12 on the Double Scale B, to $27\frac{1}{2}$ on the Double Scale A, then againft $15\frac{1}{4}$ Feet on the Double Scale B, is 35 Feet, the Content on the Double Scale A.

### *Decimally.*

*The Rule.* Multiply the Length in Feet by the Breadth in Inches, and cut off fo many Figures from the Right-hand, as are Decimals in your Length and Breadth, the Remainder divided by 12, the Quotient is the Feet, and the Remainder (if any) the odd Inches.

*Example,*

*Example,* Length is —— Feet 15.25
Breadth is ———— Inches 27.50

$$\begin{array}{r}
76250 \\
10675 \\
3050 \\
\hline
\end{array}$$ F.In.

12)419.3750(34.11
36
——————
°59
The Content is 34 Ft. 11 Inch.  48
——————
11 Inches

The Length 15 Feet ¼, or ——— Feet 15.25
Breadth 27 Inch. ½, or 27.5 inverted is — 5.72.

$$\begin{array}{r}
3050 \\
1068 \\
76 \\
\hline
\end{array}$$

Product is 419 Inch. ,⁴⁄₆ as before ; Inches 419.4
But if the Breadth be taken in Foot-meaſure,
it's done without Diviſion as follows,

F.Par.
The Length as before, 15 Feet ¼, —— 15.25
Breadth 27 Inch. ½, or 2 Ft. ,²²⁄₅⁴ inverted 92.2

$$\begin{array}{r}
3050 \\
305 \\
137 \\
\hline
\end{array}$$

Content is 34 Ft. 11 Inch, as before, or Ft. 34.92
And by the Sliding Rule it's thus : Set 1
on the Doyble Scale B, to 22.9 on the Double
Scale

Scale A; then againſt 15.25 on the Double Scale B, is almoſt 35 Feet on the Double Scale A, the Content as before.

Or, by the *Rule of Practice*, when the Length is in Feet and Inches, and the Breadth alſo in the ſame, counting for 1 Foot in Breadth, the whole Length is the Content, and for the Inches, according to its Aliquot Parts of a Foot, or 12 Inches; as it follows, in the laſt *Example.*

A Board whoſe Length is 15 Feet 3 Inches, and Breadth 27½ Inches, or 2 Feet 3½ Inches.

| | F.Inch. |
|---|---|
| The Length ——————— | 15.03 |
| Set it down again ——————— | 15.03 |
| Divide by 4, becauſe 3 Inch. is ¼ of a Foot | 3.09¾ |
| Div. the laſt by 6, becauſe ½ In. is ⅙ of 3 In. | 0.07¼ |
| And altogether is the Content required— | 34.11½ |

In like manner the *Examples* following may be done, both by the *Foot-meaſure*, and by this *Rule of Practice*, which is worth the Learner's minding.

*More Examples.*

| | | |
|---|---|---|
| The Breadth 9 Inch | Length 13 Feet | Cont. 9 Feet ¾ |
| The Breadth 21 Inch | Length 19 Feet | Cont. 33 Feet ¼ |
| The Breadth 25 Inch | Length 29½ Feet | Cont. 61 Feet ½ |

PROB. II. *Another Way when the Dimenſions are Feet and Parts, and the Content is required in Feet and Parts, by the* Sliding Rule.

Let

Let there be a Board whose Length is 24 Feet $\frac{3}{4}$, and the Breadth 1 Foot $\frac{1}{2}$, what is the Content? *Example.*

Set 1 on the Double-Scale B, to $1\frac{1}{2}$ on the Double-scale A; then against $24\frac{1}{4}$ on the Double-scale B, is 37 Feet $\frac{1}{10}$ on the Double scale A, which is the Content.

### *More Examples.*

Breadth $3\frac{1}{4}$ Feet⎫Length 20 Feet⎧ 65 Feet
Breadth $7\frac{1}{2}$ Feet⎪Length 25 Feet⎬ $187\frac{1}{2}$ Feet
Breadth 12 Feet⎨Length 30 Feet⎫ 360 Feet
Breadth 15 Feet⎭Length 35 Feet⎩ 525 Feet

How to work this second *Problem* Decimally, as is taught *Chap.* 2. *Prob.* 2.

*Example.* F.Par.

Length ——————————————— 24.75
Breadth ————————————————— 1.50

122750
2475

*Content* is 37 Ft. or $8\frac{1}{2}$ *Inch* or *Feet.*    37.1250

And by Contraction thus, the said *Example.*

F.Par.

Length 24 Feet $\frac{3}{4}$, or ———————— 24.75
Breadth 1 Foot $\frac{1}{2}$, or 1.5, inverted is — 5.1

2475
1237

The *Content* 37 Feet 1 *Inch* $\frac{1}{2}$, or *Feet* — 37.12
*Example.*

*Examples Decimally contracted.*

| | | |
|---|---|---|
| Breadth 4.27 | Len. 06.21 | 26 $\frac{51}{100}$ or **26** feet 6 Inc. |
| Breadth 8.46 | Len. 11 32 | 9, $\frac{77}{100}$ or 95 feet 9 Inc. |
| Breadth 12 54 | Len. 15.35 | 92 $\frac{43}{100}$ or 192 feet 5 Inc. $\frac{7}{8}$ |
| Breadth 16.74 | Len. 21.42 | 358 $\frac{57}{100}$ or 358 feet 6 Inc. $\frac{7}{8}$ |

**PROB. III.** *Directions for the measuring of most Sorts of Artificers Works, and first of Glazier's Work, with the Manner of taking Dimensions.*

THE beſt way to meaſure *Glazier's Work* is by the *Sliding Rule*, and the Dimenſions are taken very exact 'even to $\frac{1}{8}$ of an Inch; they commonly agree for their Work by the Foot, whether the *Glaſs* be Old or New, Squares or Quarries. *Note*, Saſh-Windows are glazed by the Square; that is, they tell how many Squares there be in all the Lights, and then reckon what they come to at ſo much the Score or Dozen.

## PROB. IV.

*Let there be a Pane of Glaſs 29½ Inches long, and 7 Inches broad, what is the Content?*

*Example.*

By the *Sliding Rule*. Set 144 (repreſented by 1.44) on the Line B, to 7 Inches on the Line A, then againſt it 29½ on the Line B, is 1 Foot and almoſt a Half on the Line A, or *Decimally* 1 Foot, $\frac{44}{100}$,

More

## More Examples.

Breadth  $3\frac{1}{2}$ Inch ⎤ Length  $20\frac{1}{4}$ Inch ⎰ 49 parts, or 5 In. $\frac{7}{8}$
Breadth  $5\frac{1}{4}$ Inch ⎸ Length  $25\frac{1}{2}$ Inch ⎱ 92 parts, or 11 In. $\frac{1}{4}$
Breadth  $9\frac{3}{4}$ Inch ⎰ Length  $30\frac{1}{2}$ Inch ⎱ 2 feet 1 Inch. $\frac{1}{4}$
Breadth  $13\frac{1}{4}$ Inch ⎦ Length  $35$  Inch ⎭ 3 feet 2 Inches, $\frac{5}{6}$

### The foregoing Problem Arithmetically.

Multiply the Length in Inches and Parts, by the Breadth in Inches and Parts, and from the Right-hand of the Product cut off fo many Figures as the Decimals in the Length and Breadth (if there be any) and the Remainder divide by 144, the Quotient is the Feet, and if any thing remain, divide it by 12, the Quotient is Inches.

### Example.

Length in Inches and Parts —— 29.5
Breadth in Inches and Parts —— 7.0

$$144)206.50(1.5$$
$$144$$

$$12)062(5 \text{ Inches}$$
$$60$$

The Content is 1 Foot 5 Inches 02
and Remainder 2.

To perform this *Decimally*, without *Division*,
Take the Length and Breadth in Foot meafure, and then it's thus;

*Example.*

*Example.*                     F. Par.

The Length 29½ Inches, or ———————— 2.46
The Breadth 7 Inches or 0.58, inverted is —85

                                  ————
                            . .    1230
                                   197
                                  ————

The *Content* is 1 Foot 5¼ Inch. or Feet 1.427
Arter the fame manner may all the foregoing
Examples be wrought.

---

P R O B.  V.  *Directions for meafuring* Joyners
*and* Painters *Work; with the Manner of taking
their Dimenfions.*

JOyners and *Painters Work*, are generally
agreed for by the Yard; and therefore having caft up their Dimenfions, they bring the
whole Sum into Yards, by dividing the Feet
by 9.

Now in taking the Dimenfions of *Joyners*
or *Painters Work*, fuch as *Pollection*, and *Bead-
work*, you muft by a Line girt Bends and Hol-
lows, and fo bring it down to the Bottom,
after which meafure the Length of the Line by
your Rule, fetting it down for one of the Dimenfions, then meafure the Length of the former Height, and fet it down for a fecond
Dimenfion.

*Note,* That in *Joyners* Meafure, *Window-fhut-
ters* and *Doors*, are *Work* and *half*, becaufe they
are worked on both fides.                 The

The Paintings of Windows are generally agreed for by so much the *Light*, and *Cafements*, at so much a *Cafement*.

In *Joyners* and *Painters Work*, Deductions are made for all Windows and Chimneys.

### *Example.*

A *Joyner* hath wainfcotted a Room 44 Feet and a half in Compafs, 9 Feet 3 quarters in Height; what is the Content by the *Sliding-Rule*.

Set 1 on the Double Scale B, to 44 Feet $\frac{1}{2}$ on the Double Scale A, then againft 9 Feet three quarters on the Double Scale B, is 433 Feet $\frac{9}{10}$ on the Double Scale A, the Content: Or thus, fet 9 on the Line B, to 44 $\frac{1}{2}$ on the Line A, then againft 9 $\frac{1}{4}$ on the Line B, is 48 Yards on the Line A, which is the Content.

*Decimally.*           F. Par

The Compafs of the Room   44.50
The Height ————————   9.75

```
                        22250
                        31150
                        40050
                        ——————— yds. feet in.
                    9)433.8750(48  01  10½
                        ·36
```

The *Content* is 48 *Yards*   073
    1 Foot 10 Inches $\frac{1}{2}$   .72
                        ——————
                        01

And

And by *Contraction*, the faid *Example* is thus,

F. Par.

The Compafs of the Room 44½ Feet is    44.5
The Height 9 Feet ¼ or 9.75 inverted is    57.9

_____

40.05
311
22

_____

The Product is ——— ——— ——— Feet    433.8
The Feet multiplied by 0.111 inverted is    111.0

_____

4338
434
43

_____

The Content as before in Yards is ——— 48.15


*How to meafure Painters Work by the* Sliding-
Rule.


*Example.* There is a piece of Painting 13
Feet ½ broad, and 23 ½ Feet long. Set 9
on the Double Scale B, to 13 ½ on the Double
Scale A, then againft 23 ½ on the Double Scale
B, is 35 Yards ¼ on the Double Scale A,
which is the Content of fuch a Piece of
Painting.


*Decimally*

*Decimally.* F.Par

Length   2 3½ Feet or —— 23.50
Breadth   1 3½ Feet or —— 13.50

$$\begin{array}{r} 117500 \\ 7050 \\ 2350 \end{array}$$

——————— *yds.Ft. In.*
9)317.2500(35 : 2 : 3
27
—————

047
Content is 3 5 Yards    45
—————
2 Feet 3 Inches    02 feet

### More Examples.

| | | | |
|---|---|---|---|
| Breadth   4 ½ feet ⎫ | Length 22   ft. ⎧ | 1 1 Yards |
| Breadth 15 ¼ feet ⎪ | Length 19 ½ ft. ⎬ | 33 Yards |
| Breadth 10   feet ⎪ | Length 30   ft. ⎬ | 33 Yds $\frac{5}{10}$ |
| Breadth 19   feet ⎭ | Length 23   ft. ⎩ | 61 Yds ¼ |

PROB VI. *The measuring of Plaisterers and Painters Work by the* Sliding Rule.

*Example 1. Of Plaistering.*

LET there be a Ceiling plaistered, whose Length is 25 Feet 3 Inches, and the Breadth 15 Feet 6 Inches : Set 9 on the Double Scale B, to 25 ¼ on the Double Scale A ; then against 15½ on the Double Scale B, is 43 Yards and almost a half, on the Double Scale A, which is the Content.     *Example*

*Example* 2. *Of Paving.*

Let there be a Piece of Paving $34\frac{1}{2}$ Feet long, and $12\frac{1}{2}$ Feet broad; set 9 on B, to $34\frac{1}{2}$ on A, then againſt $12\frac{1}{2}$ on B, is almoſt 48 Yards the Content.

## P R O B. VII.

THE meaſuring ſuch Superficies as are meaſured by the Square of 10 Feet *(viz.* 100 Feet in a Square; and if you multiply 10 by 10 it makes 100) ſuch as *Flooring, Roofing, Partitioning, Tyling,* &c.

### *By the Sliding-Rule.*

The *Rule.* Set 100, or 1 in the middle, to the Length; againſt the Breadth is the *Content.*

### *Example.*

There is a Floor whoſe Length is 61' Feet and a half, the Breadth 14 Feet and a quarter, ſet 100 on B, to $61\frac{1}{2}$ on A; then againſt $14\frac{1}{4}$ on the *Double-Scale* B, is 8 Square $\frac{3}{4}$ on the *Double-Scale* A, for the Content being 8 Square 75 Feet.

### *Decimally.*

Multiply the Length by the Breadth, and from the Produ&t cut off two more Decimals than is in the Length and Breadth, the Figures remaining are the Squares.

C

*Example.*                    F. Par.

Length 61½ Feet, or in Foot-meafure — 61.50
Breadth 14¼ Feet, or in Foot-meafure — 14.25

$$\begin{array}{r}30750\\12300\\24600\\6150\end{array}$$

Content 8¾ Squares, or 76 Feet—— 8.763750

By *Contraction,* the fame *Example* is thus,

Breadth 14¼ Feet is in Foot-meafure Feet 14.25
Length 61½ Feet, or 61.5 inverted is — 5.16

$$\begin{array}{r}8550\\142\\71\end{array}$$

Product (cutting off 3 places is) Square 8.763

*More Examples.*            *Contents.*

Length 42 feet ⎧ Breadth 24½ feet ⎫ 10 fquares 29 feet
Length 64 feet ⎨ Breadth 29 feet ⎬ 13 fquares 56 feet
Length 86 feet ⎩ Breadth 35½ feet ⎭ 30 fquares 53 feet

*Deductions in this Problem.*

In *Carpenters Work,* you are to deduct for all *Well-holes* in your *Flooring,* for all *Chimney-hearths,* and the like.

Alfo in *Partitioning,* Deductions muft be made for all *Doors* and *Windows* that are meafured in.

In *Roofing* make no Deductions for *Window-Shafts,* or *Sky-Lights,* becaufe they are more Trouble to the Workman than the Stuff is worth that would cover them.

C H A P.

*Having in the former Part of this Treatise shewed the measuring of all Manner of Superficies, likewise all Sorts of regular Figures; I shall now give you some Instructions for the taking the Dimensions of Trees, or round Timber, and so proceed to the measuring of such, and all Manner of Solids.*

THE Tree being cut down, the Custom is to Girt it, as is shewed in the begining of the next *Chapter*, and for the Length they account from the But-end, up so far as the Tree will hold half a Foot Girt, when twice folded.

The Dimensions being taken, the Tree is to be measured by *Chapter* the sixth, as square Timber.

If the Tree have great Boughs that will hold half a Foot Girt, such Boughs are called *Timber*, and they are measured and are added to the Whole.

1. *Note*, If round, rough *Timber* be measured for Sale, the common Way among Artificers in allowing for Rind or Bark is thus: If the fourth part of the Girt or Circumference of the Tree be $\frac{50}{100}$ or half a Foot, they allow $\frac{50}{100}$ or half Inch half quarter. If the fourth Part of the Circumference

be a Foot, they allow $,\frac{5.2}{0.0}$ or one Inch and almoſt a quarter; if one Foot $\frac{1}{4}$ Girt $,\frac{1.5}{0.0}$ or an Inch and above three-quarters for the Bark, &c. But for *Beach, Elm, Aſh,* and ſuch that is thin bark'd, then the Allowance muſt be a ſmall matter leſs.

2. I have ſeen great Difference in the Girt of a Tree in the Space of two Feet, or leſs, and that hath been generally where one or two Arms have been cut off: In ſuch a Caſe it is neceſſary to Girt the Tree twice or thrice, if there be any great Difference, or otherwiſe there will be loſs to the Buyer or Seller. Again, they ſay, the Buyer hath privilege to Girt any where between the Middle and the Ground End, if it be for his Advantage.

*Laſtly,* The Conſent of any Piece of *Timber* being found in Feet, if divided by 50, you have the *Content* in Loads: But ſome will have a Load to be 40 Solid Feet, therefore you may take which of the two is moſt cuſtomary with you. The Reaſon why the Difference is, they ſay, becauſe it is ſuppoſed that 40 Feet of round Timber, or 50 Feet of hewn Timber, weigh about a Tun, or twenty Hundred Weight, which is commonly accounted a Cart-Load.

C H A P.

**C H A P. VI.** *The measuring of* Round Timber *the common Way.*

TAKE the Length in Feet, Half-Feet, and if defired in Quarters, then meafure half way back again, where Girt the Tree with a fmall Chord or Chalk-line : Double this Line twice very even. This fourth-part of the Girt or Circumference meafure in Inches, Halves, and Quarters of Inches ; but be fure the Length be given in Feet, &c. and the Sides of the Square, or one-fourth of the Girt in Inches.

So you have always three Numbers given, to find the fourth, *viz.* 12 on the Girt-line for the firft, the Length in Feet always for the fecond, and the Side of the Square for the third, in Inches, Halves and Quarters.

Now we come to the *Rule.* Set 12 on the Girt-line D, to the Length (in Feet, &c.) on the Double Scale C, then againft the Side of the Square on the Girt line D, in Inches, Halves and Quarters you will find on the Double Scale C, the Content.

*Note,* This Rule is general.

*Example* 1. Suppofe the Girt of a Tree in the Middle be 64 Inches, and the Length 31 Feet, what is the Content ? Set 12 on the Girt-line D, to 31 Feet on the Double Scale C ;

C 3                              then

then againſt 16, the one-fourth of 64, on th Girt-line D, is 55 Feet, the Content on the Double Scale C.

2. A Piece of Timber is 15 Feet long, and one-fourth of the Girt 42 Inches: Set 12 on the Girt-line D, to 15 on the ſecond Length of the Double Scale C ; then againſt 42 at the Beginning of the Girt-line D, is on the Double Scale C, 184 Feet the Content, reckoning the firſt 1 at the Beginning of the Line of *Numbers* (or Double-ſcale) to be 100, the 8 Grand Diviſions 80, the two ſmall ones 4 ; ſo accounting all together makes (as I ſaid before) 184 Feet.

3. The Length is 9 Feet three quarters, and one-fourth of the Girt 39 Inches, ſet 12 on the Girt-line D to 9¾ on the Line of Numbers C, and againſt 39 at the Beginning of the Girt-line D, is 103 Feet, the Content of the Double-ſcale of Numbers C.

4. The Length 9 Inches; the Quarter of the Girt is 35 Inches. Now becauſe the Length being not one Foot, meaſure it by your Line of Foot Meaſure ; and ſee what Part of a Foot it makes, a Foot ſuppoſed to be divided into a hundred Parts : or do thus, ſet 12 on the Double-ſcale B, to 100 on the Double-ſcale A ; then againſt the Length in Inches, namely, 9 on the Double-line B, is 75 on the Double-ſcale A, for the Decimal or Part
of

of a Foot : Then set 12 on the Girt-line D, to
75 in the first Length of the *Double line* C, then
against 35 on the Girt-line D, is 6 Feet ,⁴⁄₅, or
almost half a Foot on the *Double-scale* C, for the
*Content*.

5. A *Rail* is 16 Feet long, the Quarter of
the Girt 3 Inches; set 12 on the Girt-line D,
to 16 on the first Length of the *Double-Scale* C,
then against 30 (now called 3) on the Girt-line
D, is just 1 Foot the Content, on the Double-
line C.

### More Examples.

| One-fourth | | The Length in Ft. | | The Content in Ft. and Parts are | |
|---|---|---|---|---|---|
| One-fourth | 11 Inches | | 40¼ | | 34 Feet |
| of the Girt | 16 Inches | | 39½ | | 70 Feet ₂⁄₁₀ |
| or Side of | 14 Inches ¼ | | 50 | | 68 Feet |
| the Square | 21 Inches | | 48 | | 147 Feet ₄⁄₁₀ |
| in these Ex- | 8 Inches ¾ | | 26½ | | 19 Feet ₄⁄₁₀ |
| amples is | 31 Inches | | 24 | | 160 Feet |

In taking all the Dimensions in *Foot-measure*,
it's performed by the *Sliding-Rule* sooner than
by *Inches*, thus count 10, 20, 30, 40, &c. on
the *Girt-line* to be 1, 2, 3, 4, &c. in *Foot-
measure*, and then place 10 on the *Girt-line* D,
to the Length of the Tree, on the *Double-line*
C ; then against the Girt in *Foot-measure* on the
*Girt-line* D, standeth the *Content* on the *Double-
line* C.

1. *Example* before-mentioned. A Tree in
Length 31 Feet, and Girt 64 Inches, or *Feet*
5 ,³²⁄₁₀₀, it's fourth is Feet 1 ,³²⁄₁₀₀, what's the
*Content ?*          C 4          Set

Set 10 on the *Girt-line* D, to 31 Feet on the *Double-line* C; then againſt 1 Foot 33 Parts, on the *Girt-line* D, is 55 Feet on the *Double-line* C, which is the Content as before.

2. A Piece of Timber 15 Feet long, and one-fourth of the Girt 42 Inches, or Feet 3 ,$\frac{5}{10}$, what's the Content?

Set 10 on the *Girt-line* D, to 15 on the firſt Part of the *Double-line* C, then againſt 3 Feet 5 Tenths, on the Girt-line D, is 184 Feet on the *Double line* C, the Content required. Again,

3. A Length is 9$\frac{3}{4}$ Feet, or 9 Feet ,$\frac{75}{100}$, and one-fourth of the Girt 39 Inches, or 3 Feet ,$\frac{25}{100}$: Set 10 on the *Girt-line*, to 9 ,$\frac{75}{100}$ on the *Double-line* C, and againſt 3 ,$\frac{25}{100}$ on the *Girt-line* D, is beyond 100 on the *Double-line* C: In ſuch Caſes take half the Length, and then the Content muſt be doubled. As here.

Set 10 on the *Girt-line*, to 4 ,$\frac{87}{100}$ the half of 9 ,$\frac{75}{100}$, and then againſt 3 ,$\frac{25}{100}$ is 51 ,$\frac{5}{10}$, the double is 103 Feet, the Content required.

4. The Length 9 Inches, or ,$\frac{75}{100}$ of a Foot, the Quarter of the Girt 35 Inches or 2 Feet ,$\frac{92}{100}$, ſet 10 on the *Girt-line*, to ,$\frac{75}{100}$ in the *Double-line* C, and againſt 2 Feet ,$\frac{92}{100}$ in the *Girt-line*, is 6 Feet ,$\frac{4}{10}$ on the *Double-line* C, the Content required. Again,

5. A

5. A *Rail* 16 Feet long, the Quarter of the Girt 3 Inches, or $,\frac{25}{100}$ of a Foot, what's the Content?

Set 10 on the Girt-line, to 16 in the Double-line, then against $,\frac{25}{100}$ in the Girt-line, is one Foot, the Content on the Double-line C.

And in like manner may all the *Examples* aforegoing, be wrought by the *Sliding-Rule,* if you take the fourth of the Girt in *Foot-meafure*; for 11 Inches count $\frac{92}{100}$, for 16 Inches count 1 $\frac{33}{100}$, for 14 Inches count 1 $\frac{17}{100}$, for 21 Inches count 1 $\frac{75}{100}$, for 8 $\frac{3}{4}$ Inches count $\frac{73}{100}$, and for 31 Inches count 2 $\frac{58}{100}$, and so for the Length 40 $\frac{1}{2}$ Feet is 40 $\frac{5}{10}$, and so on for any other, which may be seen by the Table of *Foot-meafure,* or by the Line of *Foot-meafure* on the *Sliding-Rule.*

If it be required to find the Content of any Piece of Timber in Loads, at 40 Feet the Load, ufe half the Girt inftead of the whole.

*Example.*

A Length is 15 Feet, and a Quarter of the Girt is 42 Inches; fet 12 on the Girt-line D, to 15 on the firft Length of the *Double-fcale* C, then against 21, (which is half of 42, one Quarter of the Girt) on the Girt-line D, is 46 on the *Double-fcale* C, whereof 4 is 4 Loads, and the 6 multiplied by 4, makes 24 Feet, which is 4 Load 24 Feet, or 4 Loads $\frac{6}{10}$.

If

If at any time you have the Contents in Loads, and you would have it in Feet; 'tis but multiplying the Content found in Loads by 4, the Square of 2, by which you divided your Girt. So 46 multiplied by 4 is 184 Feet.

If you take the Dimenfions in *Foot-meafure*, you may find the Loads contained in any Timber, thus. If a Length be 15 Feet, and quarter of the Girt 42 Inches, that is 3 Feet $\frac{5}{10}$, whofe half is 1 Foot $\frac{75}{100}$ then fet 10 on the Girt-line D, to the Length 15 on the Double-line C, then againft $\frac{75}{100}$ (the half of 3 $\frac{5}{10}$ one quarter of the Girt (on the Girt-line is 4 $\frac{6}{10}$ on the Double-line C, which is 4 Loads $\frac{6}{10}$, or 4 Loads 24 Feet, the Content required.

If at any Time you would know the Content of any Piece of Timber by Natural Arithmetic; having Girt your Tree, and taken one-fourth Part for the fide of the Square, obferve the following *Rule*.

Multiply the Length of the Side of the Square in Inches into its felf, and that produced by the Length in Feet, and the laft Product divided by 144, the Quotient is the Content in Feet, and if any thing remain, divide again by 12, the Quotient is the odd Inches.

*So*

*So in Example the Second.*

A Piece of *Timber* is 15 *Feet long*, the quarter of the Girt is 42 Inches, what is the Content of the Tree ?

The Side of the Square ————— 42 Inches
Multiplied by it felf ————— 42
                                                  84
                                              168
The Product is Inches ————— 1764
Multiplied by the Length ——— 15 feet
                                              8820
                                              1764  (feet inc.
        . 144)26460(183.9
              144 ˙˙
            ————
            1206
            1152
            ————
            00540
              432
            ————
            12)108(9 Inch.
              108
Content is 183 Feet 9 Inches—000

But by *Decimal Arithmetick*, taking all the Dimenfions in *Foot-meafure*, it's much fhorter, as follows by the forefaid *Example*.

Side

F. Par.

Side of the Square 42 Inch in *Foot meafure*  3.5
Multiply it by it's felf ——— ——— —— — 3.5

$$\begin{array}{r} 175 \\ 105 \\ \hline \end{array}$$

The Product is ——— ————————Feet  12 25
Multiply it by it's Length ———Feet — 15

$$\begin{array}{r} 6125 \\ 1225 \\ \hline \end{array}$$

Laft Product is the *Content* in *Feet* ——— 183.75
That is 183 *Feet* $\frac{75}{100}$, or 9 Inches as before.

## CHAP. VII.

*Of the Meafuring of* Round Timber or Trees, *the true Way by the* Sliding Rule *and* Arithmetically.

THE laft Way is the common Way of meafuring *Round Timber*, but it giveth not the true Content, for it is always too little, (though it be generally ufed : I fhall now give you a Point on the *Rule*, which muft be ufed inftead of 12; which Point you may cut with a fharp pointed Penknife at 10.635, but much better by the *Inftrument-maker*.
*So in the fecond Example foregoing.*
Let the Length be 15 Feet, the one fourth of the Girt is 42 Inches : Set the faid Point (which is 10.635 the true Point) to 15 the
Length

Length, then againſt 42, at the Beginning of the Girt line, is 233 Feet on the Double-ſcale for the Content ; whereas the common Way giveth but 184 Feet.

*Note,* That the common meaſure is to the true meaſure as 11 is to 14 ; ſo that if you ſet 11 on the *Double-ſcale* to the Content found the common Way, 14 ſhall point out the true Content of the ſame ; and if you ſet 14 to any true Content, 11 ſhall point out the Content the common Way ; this is done on the *Double-line.*

The Girt being taken in *Foot-meaſure,* the point for true Meaſuring is 8862 or $\frac{8862}{10000}$ or more briefly $\frac{80}{100}$, and then for the foregoing *Example.*

The Length being 15 Feet, the one fourth of the Girt 42 Inches ; that is 3 Feet $\frac{5}{10}$ or Feet 3.5 Tenths. I demand the true *Content.*

Set the aforeſaid Point $\frac{80}{100}$, on the *Girt-line* D, to the Length 15 Feet on the *Double line* C,) in its firſt Part or Beginning) ; then againſt Feet 3.5 Tenths (which is 35) on the *Girt-line* D, is 233 Feet on the *Double-line* C which is the true Content required. Again,

Suppoſe a round Tree in Length 31 Feet, and one quarter of it's Girt 16 Inches, that is 1 Foot $\frac{33}{100}$, or Foot 1.33 Parts ; what's the true Content ?

Se

Set 89 on the *Girt line* D, to 31 Feet on
the *Double line* C, then 1.33 on the *Girt-line*
D, points to 70 Feet on the *Double line* C,
the true Content required; whereas by the firſt
*Example* of *Chapter* the 6th, the Content of
this Tree is but 55 Feet, by the common Way
of meaſuring, which is no leſs than 16 Feet
ſhort of the truth.

I ſhall next ſhew how to find the true *Content*
of *Timber*, *Arithmetically*, as followeth.

By a *fixed Number*, or *Decimal*, .2821,
which multiply by the Girt of the Tree taken
in Inches, and from the Product cut off four
Figures to the Right-hand, the remaining to
the Left-hand, are the Inches for the Side of
the Square equal to the Girt or Circumference
of the Tree.

And for underſtanding it, take theſe two or
three Examples, how to find the Side of the
Square that ſhall be equal to the Girt or
Circumference of the Tree, by this *fixt Number* .2821, always to be uſed; when the Girt
is Inches.

This *Decimal*, or *fixt Number* is   .2821
The Girt in the 2d *Example* is ——   168

```
22568
16926
 2821
```

The Product is —— Inches 47.3928

Here

Here the Side of a Square equal to that *Gir*
is 47 Inches and a quarter ; for thefe four De
cimals cut off. are fo many parts of 10000 .
and .3928 is almoft ,4 of a Foot, which is
more than a quarter of an Inch.

*Another Example.*

The fixt Number is ——————— .2821
The Girt is —— ——— ——— ——— 48 Inches
                              ————————
                              22568
The Side of a Square equal —— 11284
To this Girt is 13½ Inc. or Inc.  13.5408

*Another Example.*

The fixt Number is ——— ———.2821
The Girt is —— —— ——— —— 54 Inches
                            ————————
                            11284
The Side of a Square equal —— 14105
To this Girt is 15¼ Inc. or Inc. 15.2334

I now proceed to work the foregoing piece
of Timber or Tree.

*Arithmetically.*

Multiply the Inches of the Square (with
the firft Left-hand Figure of the four which is
cut off) into its felf, and that Product by the
Length in Feet, and divide the laft Product
by 144, the Quotient is Feet, and the Remainder, if any, divide by 12, the Quotient is
Inches.

*Exam-*

*Example.*     F. Par.

The fide of the Square in Inches — 47.3
Multiply by itfelf ——————— 47.3

$$1419$$
$$3311$$
$$1892$$

In the Product cut off 2 Decimals 2237.29 Pt.
The Length in Feet ——————— 15

$$11185$$
$$2237 \qquad \text{F. I.}$$

$$144)33555 \backslash 233.3$$
$$288 \cdot\cdot$$

$$0475$$
$$432$$

$$0435$$
$$432$$

The Content of this *Timber* ⎱
is 233 Feet 3 Inches — ⎰ — 003 Inch

*Another Example for Practice.*

There is a *round Tree*, whofe *Girt* is 48 *Inches*, and the *Length* 8 Feet 9 Inches, what is the Content?

Multiply the Inches of the Side of the Square equal to the Girt with the firft Figure on the Left-hand, of the four cut off) into it felf, and from the Product cut off two Figures to the Right-hand, the Remainder multiply by the Inches of the Length of the Tree (and not by the Length in Feet) becaufe the Length is Feet
and

and Inches, therefore you muſt bring your Length into Inches by multiplying them by 12, this laſt Product divide by the Inches that are in a ſquare Foot of Timber, namely 1728, the Quotient is the Feet, the Remainder (if any) divide by 144, the Quotient is the odd Inches.

*Example.*

The Side of the Square being Inches — 13.5
Multiply it by itſelf —————————13.5

|  | F.In. | 675 |
| The Length —— | 8 9 | 405 |
| Multiplied by — | 12 | 135 |
| Makes ———— | 105 Inches —— | 182.25 |
| The Length in Inches is———— | 105 In. | |

                                    910
                                  1820
The Content of this ⎫   1728)19110(11 Ft.
round Tree is 11 Feet, ⎪       1728
and 102 remains, which ⎬      01830
is not one Inch. ⎭            1728
                            0102

*Note,* The fix'd Number .2821 is the Side of a Square, that's equal to a Circle whoſe Girt is 1 Inch, 1 Foot, or one any Thing.

Alſo the Number 10.635, is the Side of a Square that's equal to a Circle whoſe Diameter is 12 Inches.

D                                    Alſo

Alſo .8862, (or ſhorter 89) is the Side of a Square that's equal to a circle whoſe Diameter is 1 Foot, or one any Thing.

Now to perform this Arithmetically, when all the Dimenſions are taken in Foot-meaſure.

The *Rule.* Multiply the whole Girt by the fix'd Number .2821, and the Product is the Side of a Square equal to that Girt; then multiply that Product by itſelf, and this laſt Product multiply by the Length of the Tree, and it produceth the Content required.

*Example.* Let the Length be 15 Feet, and the Girt 14 Feet. I demand the true Content?

| | |
|---|---:|
| The fix'd Number is —————— ——— | .2821 |
| Multiply by the Girt 14 Feet, inverted is | 41 |
| | 2821 |
| | 1128 |
| The Product is the Side of the Square Ft. | 3.949 |
| Multiply by itſelf, inverted is ——— | 040.3 |
| | 11.847 |
| | 3546 |
| | 156 |
| | 27 |
| The Product is the Square ——— Feet | 15.576 |
| Multiply by the Length 15 Feet, inverted | 51 |
| | 15.576 |
| | 7785 |
| The Product is the Content —— Feet | 233.61 |

That is 233 Feet $\frac{61}{100}$, or 233 Feet 7 Inches.

*Another*

*Another Example for Practice.*

A round Tree, whose Girt is 48 Inches, or 4
Feet, and Length 8 Feet 9 Inc. or 8 Ft. $\frac{75}{100}$.
The fix'd Number is ——— ——— ——— .282
Multiply by the Girt 4 Feet ——— ——— 4
Product is the Side of the Square—Foot 1.128
Multiply by itself, inverted is ——— ——— 8.211

$$\begin{array}{r} 1128 \\ 112 \\ 22 \\ 8 \\ \hline \end{array}$$

The Product is the Square ——— Foot 1.270
Multiply by the Length 8.75, inverted is  5.78

$$\begin{array}{r} 10160 \\ 889 \\ 60 \\ \hline \end{array}$$

The Product is the Content ——— Feet 11.109

# C H A P.  VIII.

### *To measure the* C U B E.

LET there be a Cube whose Sides are 3 Feet
3 Inches, what is the Content?

### *By the Sliding-Rule.*

Set 12 on the Girt-line D, to 3¼ on the
Double-scale C ; then against 39 Inches the Side
of the Cube on the Girt-line D, is 34 Feet and ¼
on the Double-scale C, the Content required.

*De-*

*Decimally.*

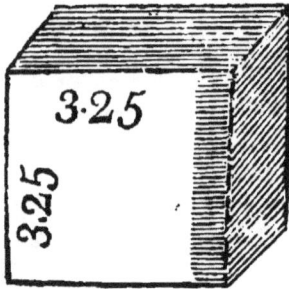

3·25
3.25

F.Par.
The Side 3 ft. 3 in. or 3.25
Multipl. by the fame 3.25
—————
1625
650
975
—————
The *1ft Prod.* is  10.5625
Multipl. by the fame 3.25
—————
528125
211250
316875
—————

The Content is 34 ⅓ Feet or — Feet 34.328125

*But by Contraction thus,*

F.Par.
The Side of the Cube 3 Feet 3 Inches or  3.25
Multiply by 3 Feet 25 Parts, inverted is  52.3
—————
975
.  65
16
—————

The firft Product called the Square, is Ft. 1056
Multiply by 3 Feet, 25 Parts inverted is—52.3
—————
3168
211
53
—————

The Content 34 ¦ Feet as before——Ft. 34.32
C H A P.

## C H A P.  IX.

*To meafure Timber that is neither Round, nor
Square; but firft to find a Mean Proportional
between any two Numbers.*

TO find this *Mean Proportional*, fet the
greater of the two Numbers on the
Girt-line, to the fame on the Double-line of
Numbers; then againft the leffer on the Double-
line of Numbers, is the Mean Proportional
on the Girt-line; or fet the leffer on the Girt-
line, to the fame on the Double-line of Numbers; then againft the greater on the Double-
line of Numbers, is the Mean Proportional on
the Girt-line; any one of thefe will do.

*Note,* Examples of this are in the next *Problem.*

### P R O B.  I.

*To meafure unequal fquared Timber, that is, when
the Breadth and Depth are not equal.*

Meafure the Length of the Piece in Feet,
and the Breadth and Depth, (at the End) in
Inches. Then find the Mean Proportional
between the Breadth and Depth of the Piece
as is taught above, and in the Example
following. The Mean Proportional is the Side
of a Square equal to the End of the Piece;
which having found; the Piece may be meafured as Square Timber.

*Ex-*

## Example I.

In a Piece of Timber whofe Length is 13 Feet, the Breadth 23 Inches, and the Depth 13 Inches ; fet 23 on the Girt-line D, to 23 on the Double-fcale C ; then againft 13 on the Line C is 17.35, or 17 ⅓ on the Girt-line D, then fetting 12 on the Girt-line D, to 13 Feet the Length on the Line of Numbers C, then againft 17 ⅓ (the Mean Proportion) on the Girt-line D, is 27 Feet the Content required.

## Example II.

In Stone, which let it be 7 Feet 40 Parts, or 7 Feet and about 5 Inches in Length, and 30 Inches in Breadth, and 23½ deep ; fet 30 Inches on the Girt-line D, to 30 on the Double-fcale C ; then againft 23½ on the Line of Numbers C, is 26½, on 26.50 on the Girt-line D : Then fet 12 on the Girt-line D, to 7.40, on the Line C, then againft 26½ on the Girt-line D, is 36 feet the Content on the Double-fcale C.

The Content of thefe two Examples by Arithmetic, is thus.

## The firft Example.

Let there be a Piece of Timber whofe Length is 13 Feet, the Breadth 23 Inches, and the Depth 13 Inches.

*The*

## The Rule.

Multiply the Breadth in Inches, by the Depth in Inches; and that Product multiply by the Length in Feet, and divide the laſt Product by 144, the Quotient is Feet, and the Remainder (if any) divide by 12, the Quotient is Inches.

## Example.

The Breadth    23 Inches.
The Depth — 13 Inches.

         69
         23     *Note,* This Example by the
The Product 299     *Sliding Rule* was 27 Feet,
Multiply by   13 Ft.    but by Arithmetick it
         897      wants ſomething; for it
         299      is 26 Feet 11 Inches, and
144)3887(26 Ft. $\frac{11}{12}$ of an Inch.
     288

     1007
      864

   12)143(11 Inches
      12

     023
     12

    11 Remainder is $\frac{11}{12}$ of an Inch.

*By Foot-meafure it's done thus :*

F.Par.

Breadth 23 Inches in Foot-meafure is———— 1.92

Depth 13 Inches or Feet 1.08, inverted is 80.1

---

1.92

15

---

The Product is ——————————— Feet —— 2.07

Multiply by the Length 13 Feet, inverted is 31

---

207

62

---

Content is 26 Feet 11 Inches, or Feet ——— 26.9

*The Second Example, Decimally.*

Let there be a Piece of Stone 7 Feet 40 Parts long, 2 Feet 50 Parts in Breadth, and 1 Foot 96 Parts in Depth, what is the Content ?

*The Rule.*

Multiply the Length by the Breadth, and the Product by the Depth or Thicknefs ; and cut off from the laft Product, as many Places to the Right hand as there are Decimals in the three Dimenfions, the Integers remaining are Feet.

*Example.*

*Example.*

F.Par.

The Length ⸺⸺⸺⸺⸺⸺ 7.40
The Breadth ⸺⸺⸺⸺⸺⸺ 2.50
_____

37000
1480
_____

The Prod. Feet 18.5000
The Depth Ft.    1.96

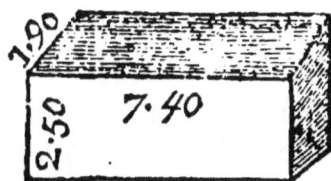

1110000
1665
185
_____

The Content is 36 Ft. 3 Inc. or Ft.  36.260000

*By Foot-meaſure it's thus contraƈted.*

F.Par.

The Stones Length 7 Feet 5 Inches, that is 7.41
Stone Breadth 30 Inches or 2.5 inverted is  5.2
_____

1482
370
_____

The Produƈt is ⸺⸺⸺⸺⸺Feet 18.52
Depth 23½ Inches, or 1.96, inverted is —  69.1
_____

1852
1667
111
_____

The Content is 36 Feet 4 Inches, or Feet 36.30

PROB.

## P R O B. II.

*To find the Content of a Piece of Timber whose End is in the Form of a Triangle ; and both Ends alike and equal.*

TO find the Content of such Timber, first find a mean Proportional between the Base and half the Perpendicular of the Triangular End, or between the Perpendicular and half the Base, both measured in Inches. Then is the mean Proportional the Side of a Square equal to the Triangle.

Then to find the Content, set 12 on the Girtline D, to the Length in Feet on the *Line of Numbers* C, then against the mean Proportional on the Girt line D, is the Content on the *Line of Numbers* C.

But the Dimensions being all taken in *Foot-measure*, and the mean Proportional found in the same ; then set 1 on the Girt-line to D, the Length in the Double-line C ; then against the mean Proportional on the Girt line D,·is the Content in the Double-line C.

*Note*, If the two Sides of a Triangle be equal, the other is Side is called the Base, but if the three Sides be unequal, the longest Side is the Base : From whence the nearest Distance, to the opposite Angle, is the Perpendicular.

*Example.*

*Example.* A Piece of Timber 19 Feet 6 Inches
in Length, the Base of the Triangle at each End
being 21 Inches, and the Perpendicular to each
Base being 16 Inches, what's the Content?

Set 21 Inches on the Girt-line D, to 21 on
the Double-line C; then against 8 on the Line
C, is 12 ?, or 12.95 on the Girt-line D, the
Mean Proportional.

Then set 12 on the Girt-line D, to 19½ Feet
the Length on the Double-line C; and against
12.95 (the Mean Proportional) on the Girt-line
D, is 22 ¾, or Feet 22 ⁸⁄₁₀ (the Content required)
on the Double-line C. Or thus, take all the
Dimensions in Foot-measure, and then the Length
19 Feet 6 Inches is 19.5, the Base 21 Inches is
1.75, and the Perpendicular 16 Inches is 1.33;
now set 1.75 on the Girt-line D, to 1.75 on the
Double-line C, and against 0.67 on the Double-
line C, is 1.08 on the Girt-line D, for the Mean
Proportional: Then set 1 on the Girt-line D, to
the Length 19.5 on the Double line C, and
against 1.08 on the Girt line D, is 22.8 (or
Feet 22 almost 11 Inches) on the Double-line
C, for the Content, very near to that before.

*Arithmetically.*

The *Rule.* Multiply the Base by the Perpendicular (both taken in Foot-measure) and
half that Product multiply'd by the Length,
                              produceth

produceth the Content required, as in the forefaid Example.

F.Par.

The Bafe 21 Inch, or in Foot meafure is— 1.75
The Perpendicular 16 Inch, or 1.33 inver. 33.1

$$\overline{\phantom{xxxx}}$$

1.75
53
5

$$\overline{\phantom{xxxx}}$$

The Product is—————————————Feet 2.33

Half the Product is————— —————Feet 1.17
The Length 19 Ft. 6 Inch. 19.5 inver. — 5.91

1:7
105
6

$$\overline{\phantom{xxxx}}$$

The Content is 22 Feet 10 Inches, or Ft. 22.8

PROB. III. *To meafure Timber that Tapereth.*

MEafure the Length in Feet, and note, one-third of the Length, which may be found by the *Sliding-Rule*, thus, Set 3 on the Double-line A, to the Length on the Double-line B; then againft 1 on the Double-line A, is the third part on the Double-line B; then if the Solid be round, Meafure the Diameter at each End in Inches, fubtracting the leffer Diameter from the greater, and add half the Difference to the leffer Diameter, the Sum is the Diameter in the middle of the
Piece

Piece; then fet 13.54 on the Girt-line D, to
the Length on the Double-line C, then againſt
the Diameter in the middle on the Girt-line D,
is a fourth Number on the Double-line C,
then fet 13.54, or 13½, of the Girt line D, to
the third part of the Length on the Double-line
C; then againſt the half Difference on the
Girt-line D, is another fourth Number on
the Double-line C, theſe two fourth Numbers
added together give the Content.

*Example.* Let the Length be 27 Feet (the
one third by the Rule is 9) the greater Diameter 22 Inches, and the leſſer 18 Inches.

See the Operation.

The greater ———————————— 22 Inches
The leſſer Diameter ————— 18 Inches
The Sum is——————————— 40
The Difference —————— 04
The half Difference ———— 02 ⎱ added
The leſſer Diameter ——— 18 ⎰
The Diameter in the middle——20 Inches

Then fet 13.54 or 13½ on the Girt-line D,
to 27 on the Double-line C; then againſt 20
on the Girt-line D, is 58.900, that is 58 Feet
$\frac{900}{1000}$ Parts.

Again, fet 13.54 on the Girt-line D, to 9
on the Double-line C; then againſt 2 on the
Girt-line (repreſented by 20) is 196 Parts of
a Foot.

The

$\left\{\begin{array}{l}\text{The firft fourth Number} \text{—— } 58.900 \\ \text{The fecond fourth Number} \text{— } 196 \\ \text{The Sum is the Content } \text{——} 59.096\end{array}\right\}$ added

Which is 59 Feet ¼ Inch.

*Note*, If the Timber be wainy, that is neither Square nor Round, take one Wain in and leave the other out.

The fame done by *Foot-meafure*.

| | F. Par. |
|---|---|
| The greater Diameter 22 Inches, or —— | 1.83 |
| The leffer Diameter 18 Inches, or —— | 1.50 |
| The Sum of both Diameters is — Foot | 3.33 |
| Half is the middle Diameter —— Foot | 1.67 |
| Difference of the Diameter is —— Foot | 0.33 |
| Half Difference of the Diameter is Feet— | 0.17 |

Then fet 1.13 on the Girt-line D, to the Length 27 Feet on the Double-line C; and then againft 1.67 on the Girt-line D, is 58 Feet , ⁹⁄₁₀. Then,

Set 1.13 on the Girt-line D, to 9 Feet on the Double-line C; and then againft 0.17 on the Girt-line D, is .196 Parts of a Foot, and both put togethe, is the Content; that is, 58.9 and .196 added makes 59.096, or 59 Feet ¼ Inch, as before.

Suppofe the Solid Square, and having the fame Dimenfions; that is Length 27 Feet, the Greater End 22 Inches Square, and the Leffer End 18 Inches Square, what's the Content?

The

The greater Square  22—Inches or——— 1.83
The leſſer Square    18—Inches or——— 1.5
Sum of both Squares 40—Inches or——— 3.33
½ is the mid. Square 20—Inches or——— 1.67
Differ. of the Ends   4—Inches or——— 0.33
½ Differ. of the Ends 2—Inches or ——— 0.167

Then to find the Content by the Inch-meaſure it's thus. Set 12 on the Girt-line D, to 27 Feet (the Length of the Solid) on the Double-line C ; then againſt 20 Inches (the middle Square) on the Girt-line C, is 75 Feet $\frac{4}{10}$. Again, ſet 12 on the Girt-line to 9 Feet (one third of the Length) on the Double line ; and againſt 2 Inches the half Difference of the Ends) on the Girt-line, is $\frac{25}{100}$ parts of a foot ; both together is 75 $\frac{65}{100}$ parts (which is almoſt 8 Inches) the Content of the Solid : Or thus, by Foot-meaſure.

Set 1 on the Girt line, to the Length 27 Feet on the Double-line, then againſt 1 Foot 67 Parts (the Middle Square) on the Girt-line, ſtands 75 Feet $\frac{4}{10}$, and ſetting 1 on the Girt-line, to 9 Feet (the one Third of the Length) on the Double-line ; then againſt 0.167 (the half Difference of the Ends) on the Girt-line is (on the Double-line) $\frac{25}{100}$ Parts of a Foot, and both together

ther 75 Feet $\frac{4}{18}$ and $\frac{15}{180}$ is 75 Feet $\frac{61}{180}$ part
of a Foot as before, exactly agreeing both by
Inch-meafure and Foot-meafure ; and here note,
the Numbers that are fix'd in this kind are
thus to be underftood: 13.54057, &c. in
fhort 13.54 is the Diameter of a Circle, whofe
Area or Content is 144 Square Inches: And
1,1283814, &c in fhort 1.12 is the Diameter
of a Circle whofe Area or Content is 1 Foot, or
one any Thing.

## PROB. IV.

*To find how many Inches in Length will make
a Foot Solid at any Girt, being the Side of a
Square, not exceeding 40 Inches.*

LET the Girt or Side of the Square upon
the Girt-line be fet to 1 on the Numbers,
then againft 41.57 of the Girt line is the Number of Inches on the Numbers that will make a
Solid Foot.

### Example.

Let the Side of a Square be 7 Inches, fet 7
on the Girt-line D, to 1 on the Double-line C,
then againft 41.57 on the Girt-line D, is 35 $\frac{1}{4}$
Inches on the Double-line C, for the Length of
1 Foot Solid.

To do the fame in *Foot-meafure* ; the Side
of the Square 7 Inches, in *Foot-meafure* is .58
Parts ; which faid .58 Parts on the Girt-line
D, being fet to 1, on the Double-line C ;
then

then againſt 1 on the *Girt-line* D, is 294, that is 2 Feet 94 parts, or 2 Feet 11¼ Inches for the Length, to make one Foot of *Timber*.

## ·P R O B. V.

*Having the Diameter of a Circle, or Round Piece of Timber; to find the Side of a Square within the Circle, or to know how many Inches the Side of the Square will be when the Round Timber is ſquared.*

### The Rule.

Set always 8.5 on the *Double-line* A, to 6 on the *Double-line* B; then againſt the Diameter on the Line A, is the Side of the *Square* on the Line B.

### Example.

Let the Diameter be 18 Inches; ſet 8.5 on A, to 6 on B; then againſt 18 on A, is 12¼ on the Line B, for the Side of a Square within that Circle.

The ſame done in *Foot-meaſure*, the *Diameter* being 18 Inches is in *Foot-meaſure* 1.5; Then ſet 1 on the *Double-line* A, always to .707 or .71 almoſt on the *Double-line* B; then againſt the Diameter 1.5 on the *Double-line* A, is 1.7 on the *Double-line* B; that is Foot 1.7 Tenths the Side of a Square within a Circle, whoſe Diameter is Foot 1.5 Tenths. And here note the given Numbers 8.5, and 6 (but more truly 1, and .707) are, on the Diame-

E

ter of a Circle, the other the Side of a Square within that Circle.

## PROB. VI.

*By having the Girt of a Tree, or round Piece of Timber, to find the Side of a Square within.*

### The Rule.

Set 10 to 9 on the Line A and B, then a-gainft the *Girt* on the Line A, is the Inches for the Side of the Square on the Line B.

### Example.

Let the Girt be 16 Inches, fet 10 on the Line A, to 9 on the Line B; then againft 16 on the Line A, is 14¼ on the Line B, for the Side of the Square.

By *Foot-meafure* it's thus: the Girt 16 Inches, is Foot 1.33 parts; then fet 10 on the *Double-line* A, to 9 on the *Double line* B, and againft the Girt 1.33 on the *Double-line* A, is on the *Double-line* B, 1.19; that is 1 Foot 19 Parts, which is 1 Foot 2 ¼ Inches, for the Side of the Square within.

And *Note*, the Numbers 10 and 9, or 1 and 9, fhews when the Square within the Circle is 1, the fourth part of the Circumfe-rence parts of the fame.

By

By thefe two laft Problems you may know (before a Piece of Timber is hewn) how many Boards or Planks of any Thicknefs it will make.

## PROB. VII.

*By the fourth Part of the Girt of Round Timber, to find the Side of a Square equal to it.*

### The Rule.

Set 1 on the Line A, to 1.128 on the Line B; then againſt the (one-fourth of the whole) Girt on the Line A, is on the Line B, the Side of the Square equal to it?

*Example.* Let the Girt (that is the one-fourth of the whole Girt) be 16 Inches; how much is the Side of the Square equal to it?

Set 1 to 1.128, but ſhorter to 1.13, on the Lines A and B; then againſt 16 on the Line A, is 18 on the Line B; which ſheweth that a Square whoſe Side is 18 Inches, is equal to a Circle whoſe Girt is 64 Inches, or one-fourth of its Girt is 16 Inches.

By *Foot meaſure*, the Rule being ſet as before, 1 againſt 1 13, then againſt Foot 1.33 Parts (equal to 16 Inches) you will find 1.5, that is 1 Foot $\frac{5}{10}$, equal to 1 Foot 6 Inches, the Side of a Square equal to the given Girt, the ſame as before.

E 2                    C H A P.

## CHAP. X.

### *To meaſure* Brick-Work.

**B**Rick Work is meaſured by the *Rod* of 16
Feet and a half, whoſe Square is 272 $\frac{1}{4}$
which ſheweth that the Point on the *Double-
line* is 272 $\frac{1}{4}$, but 272 being marked on the *Rule*
will ſerve (without any conſiderable Error) a
Brick and a half thick, and is the fixed Number.

### *An Example.*

There is a *Brick Wall* whoſe Length is 564 Feet
the Height is 10 Feet and $\frac{1}{2}$. Set 272 on the
*Double-line* A, to 564 on the Line B ; then
againſt 10$\frac{1}{2}$ on the Line A, is 21 $\frac{7}{10}$ *Rod,* the
Content on the *Double-line* B.

*Note* 1. Always the fix'd Number that goeth
with the Queſtion is called the firſt *Number,*
which you may ſet to either of the other
Numbers.

*Note* 2. That 272 $\frac{1}{4}$ Feet makes a *Rod* of
*Brick-work,* at a *Brick* and a half Thickneſs ; if
it be thicker fewer Feet go to a *Rod ;* if thinner
the more.

If you demand how many Feet makes a *Rod*
at two *Bricks* Thickneſs : Two being the firſt or
fixed Number that goeth with the Queſtion muſt
be ſet to 1 $\frac{1}{2}$ ; then againſt 272 $\frac{1}{4}$ is about 204,
*viz.* 204 Feet $\frac{10}{88}$, and ſo for any other Thickneſs, for which

*Take*

## *Take this General Rule.*

Set 1½ to any Thickneſs; then againſt 272 ¼ on the ſecond *Length* is the Number of *Feet* that makes a *Rod* to that Thickneſs, and is called the firſt Number for that Thickneſs, as in the *Table* following.

By this Table, made by the ſame *Rule*, by the *half Brick's* Thickneſs of the Wall, you have the firſt *Number* by Inſpection; as thus, againſt 2 *Bricks* ½ or 3 *half Bricks* thick, is 163 *Feet*, for a Square *Rod*, and ſo for any other.

| Half Bricks thick. | | Square Feet in a Rod on the Superficies. | Half Bricks thick. | | Square Feet in a Rod on the Superficies. |
|---|---|---|---|---|---|
| ½. | 1 | 817 Feet | 3 ½. | 7 | 117 Feet |
| 1. | 2 | 408 Feet | 4. | 8 | 102 Feet |
| 1 ½. | 3 | 272 Feet | 4 ½. | 9 | 91 Feet |
| 2. | 4 | 204 Feet | 5. | 10 | 82 Feet |
| 2 ½. | 5 | 163 Feet | 5 ½. | 11 | 74 Feet |
| 3. | 6 | 136 Feet | 6. | 12 | 68 Feet |

## CHAP. XI.

### *Meaſuring of* LAND.

LAND is uſually meaſured by the *Pole*, *Perch* or *Rod*, which is 16 ½ *Feet* long; 40 *Poles* in Length and 4 in Breadth makes an *Acre*, ſo that an *Acre* of *Land* contains 160 *ſquare Poles*, half an *Acre* is 80, and a quarter is 40,

Some use a *Pole-Chain*, and others a 4 *Pole Chain*, the best for Use is that of 4 *Poles*. This *Chain* is called *G U N T E R's Chain*. It is divided into 100 Links or Parts, every Link is 7 Inches and 92 Parts of an Inch in Length.

## P R O B.  I.

### By the Pole Chain.

There is a Plot of Ground 35 Pole or Perch broad, and 185 Pole long, how many Acres is the Content?

### By the Sliding Rule

Thus, Set 160 to the Breadth, then against the Length is the Content.

### Example.  –

Set 160 on the *Double-line* A, to 35 the Breadth on the *Double-line* B ; then against 185 the Length on the Line A, is $40\frac{16}{100}$ (that is 40 Acres and almost $\frac{1}{2}$) on the *Double-line* B ; the Content.

### Arithmetically.

*The Rule.* Multiply the Length in Poles by the Breadth in Poles, and divide the Product by 160 (the Poles in an Acre) the Quotient is the Acres, if any remains divide again,

By $\left\{\begin{array}{c}120\\8\\40\end{array}\right\}$ the Quotient is $\left\{\begin{array}{c}\frac{3}{4}\\\frac{1}{2}\\\frac{1}{4}\end{array}\right\}$ of an Acre.

*Example*

### Example.

The Length ———————— 185 *Poles*
The Breadth ——————— 35 *Poles*

But 75 *Pole*, remaining after the firſt Diviſon, viz. by 160, which 75 being divided by 40 the laſt Remainder is 35, which makes the *Content* 40 *Acres* 1 *Quarter* and 35 *Poles*, or almoſt 40 *Acres* and ½.

$$9^25$$
$$5 \dot{5} 5$$
$$16|0,6471 5(40 Acr.$$
$$64$$
———————
$$40|0075(1 \ quar.$$
$$4$$
———————
35 *Poles*
(remaining.

### More Examples by the Rule.

Breadth $\left.\begin{array}{c}42\\53\\63\end{array}\right\}$ Poles. $\left.\begin{array}{c}\end{array}\right.$ Length $\left.\begin{array}{c}49\\64\\85\end{array}\right\}$ Poles. $\left\{\begin{array}{l}\text{12 Acres },\frac{66}{100}\text{ or 12 }\frac{3}{4}.\\ \text{21 Acres }\frac{2}{30}\text{ or 21 }\frac{1}{4}.\\ \text{34 Acres juſt.}\end{array}\right.$

## PROB. II.

*By the four Pole Chain. There is a Plot of Ground containing 16 Chains 25 Links in Breadth, and 57 Chains 30 Links in Length, what is the Content of that Piece of Land?*

### By the Sliding Rule.

*The Rule.* Set 10 on the *Double-line*, to the Breadth in *Chains* and *Links* on the other, then againſt the Length in *Chains* and *Links* is the *Content?*

*Example*

*Example.*

Set 10 on the *Double-line* A, to 16.25 on the *Double-line* B, then againſt 57.30 on the *Double line* A, is 93 Acres and ſomething more, for the Content on the *Double-line* B.

*Arithmetically.*

Length —— Chains 57.30 Links
Breadth —— Chains 16.25 Links

$$
\begin{array}{r}
28650 \\
11460 \\
34380 \\
5730 \\
\hline
\end{array}
$$

Acres 93.11250 cut off five Figures

$$
\begin{array}{r}
4 \\
\hline
\end{array}
$$

no Roods — | 45000
        | 40

Poles —— 1800000

*Note,* 4 Rods make an *Acre,* and 40 *Poles, Perches* or *Rods* make one *Rood* or *Quarter* of an *Acre.*

*More Examples by the Rule.*

*Breadth* $\left\{\begin{array}{l} 3.35 \\ 4.53 \\ 6.28 \end{array}\right\}$ *Length* $\left\{\begin{array}{l} 15.32 \\ 16.42 \\ 72.51 \end{array}\right\}$ 5 *Acr.* a little more
7 ½ *Acres* almoſt.
118 *Acres.*

*Note,* Gunter's *Chain* containeth four *Statute Poles* in 100 *Links,* ſo that any Number of *Chains* is no more than ſo many 100 *Links* ; as 4 *Chains* is 400 *Links,* 6 *Chains* is 600 Links,

Links, &c. and 160 Square Statute Poles is an Acre, each Pole being 16 Foot and a Half. Therefore in a square Chain is 16 square Poles. If you divide 160 (the Square Poles in an Acre) by 16 (the Square Poles in a Chain) the Quotient is 10, the square Chains in an Acre.

A Square Chain contains 10000 Square Links (for it is 100 multiplied by 100) and consequently an Acre is 100000 Square Links.

Therefore the Reason of the foregoing Operation is evident, for if you multiply 5730 Links by 1625 Links, the Product will be 9311250 Square Links. Divide 9311250 by 100000 (the Square Links in an Acre) the Quotient is Acres, and the Remainder is Parts of an Acre. But to divide by 100000 is no more than to cut off five Figures to the Right-hand, so will the Remaining Figures to the Left-hand be Acres, and those cut off to the Right hand 11250, are Parts of an Acre: Again, multiply those five Figures cut off from the Right-hand of the Product by four, and from the Product cut off five Figures to the Right-hand, those on the Left-hand will be roods, and if those cut off be multiplied by 40, and from the Product five Figures be cut off from the Right-hand, those on the Left-hand, are Poles or Perches.

So

So in the preceding Operation there is 93 Acres, no Roods, 18 Perches for the Content.

## PROB. III.

*How to reduce Statute Meafure to Cuftomary, and the contrary.*

ACcording to a *Statute* made in the 33d of *Edward* the Firft, and likewife in the 24th of *Elizabeth*, a Statute Pole is 16 Feet and a Half long. Now divers Parts of *England* ufe a Pole of 18 Feet long, and fome a Pole of 21 Feet long, and others a Pole of 24 Feet long. Therefore to turn one Sort of Meafure into another; fuppofe a *Statute* into *Cuftomary*, do thus; multiply any Number of Acres, Roods and Perches *Statute Meafure*, by the Square half Yards, or Square half Feet, in a Square Pole of *Statute Meafure*, the Product divide by the Square half Yards, or Square half Feet in a Pole of the *Cuftomary Meafure*, the Quotient gives you the Acres, Roods and Perches of that *Cuftomary Meafure*; as for *Example*. In 172 Acres *Statute Meafure*, how many Acres of 18 Feet to the Pole or Perch?

In a Statute Pole or Perch, there is five Yards and a Half, that is 11 half Yards; confequently in a fquare *Statute* Pole there is 121 fquare half Yards, and in a fquare Pole of 18 Feet, there is 144 fquare half Yards,

172 Yards

172 Acres Statute Meafure
121 Square half Yards in a Statute Pole.

```
     172
     344
     172
144)20812(144 Acres Cuftomary, and 76/144.
     144
     0641
      576
      0652
       576
       076 Overplus
```

Therefore the Cuftomary Acres (at 18 Feet
to a Perch) is 144 76/144 of an Acre, the Overplus multiply by four, and the Product divide
by 144 gives Roods; again, multiply the
Remainder by 40, and the Product divide
by 144, the Quotient will be Perches; as by
the following Operation of the preceding Remainder.

76 Overplus, or Remainder.

```
        4
144)304(2 Roods
    288
    016
     40
144)640(4 Perches
    576
    064
```

So

So that 172 Acres Statute Measure is 144
Acres, 2 Roods and 4 Perches, with a small
Remainder, not worth Notice, of *Customary
Measure*, of 18 Feet to the Pole or Perch.

If there had been any given Roods and
Perches with the Acres, then you must turn
all into Perches, and having multiplied and
divided as before directed, divide the Quotient
by 160, or 40, and that Quotient again by 4

P R O B,

## P R O B.  IV.

*In* 543 *Cuſtomary Acres, at* 18 *Feet to the Perch, how many Acres of* Statute-Meaſure ?

543 Cuſtomary Acres.
144 ſquare ½ Yards in a cuſtomary *Acre.*

2172
2172
543
121)78 92(646 Statute *Acres.*
726

0559
484

0752
726

26 Remainder, multiplied
by ——  4

104 no Roods
40
121)4160(34 Perches
363

530
484

046 Remains

Here

'Here you fee that in 543 Cuftomary Acres at 18 Feet the Perch, there are 646 Acres, no Roods, 34 Perches Statute Meafure, and $\frac{46}{121}$.

Cuftomary Acres as well as Statute Acres, contain 160 Square Perches, the Difference is only in the bignefs of the Perch.

## C H A P. XII.

*The Defcription and Conftruction of Scamozzi's Lines upon the Two Foot Rule, with their Ufe in all the neceffary Problems preparatory to the Practice of Building.*

1. **T**HE moft ufeful Line of all is a Scale of equal Parts, beginning at o at the Center or Joint of the Rule, and proceeding to 30 near the remote End of it on both Sides ; at which on each Leg of the Rule there is a Brafs Point or Center-Pin fix'd, and as thefe 30 Divifions are intended in general to reprefent 30 Feet, they are every one fubdivided into 12 equal Parts, every one of which reprefents an Inch ; but they may be applied to any other Meafure.

2. Next to this, and above the 30 Scale is the 40 Scale, or the fame Length divided into 40 equal Parts, and each into 6 Subdivifions, fo that if the Divifions be Feet, the Subdivifions are each of them 2 Inches.

3. On

3. On the other Joint of the Rule, and next without the 30 Scale is a Line of Chords with a Center-Pin, at 60 equal to 15 on the 30 Line, and fo proceeding to 180, where there is another Center-pin at the end of the Line.

4. Next to the Line of Chords is a Line of Tangents, beginning at .0, and continuing to the Tangent of 45, which is the Radius of every Circle.

5. The Vacancy left by the fpreading of the 30 Scale is fupply'd by the Line of *Polygons*, it begins at 2 at the remote End of the Scale, and reckons inwards in fome Scales to 20; but it is now thought fufficient to continue it to 12, to prevent the Confufion upon the Scale which would otherwife be occafioned by the Contraction of the Lines, and the decreafing Diftance of the Numbers towards the Center.

Before I give a further Account of the Ufe of *Scammozzi's* Lines, it will be proper to explain the Terms made ufe of in Buildings, fo far as it relates to the Ufe of thofe Lines; as

1. A Model or Module is the Diameter of a Pillar or Column, and a Minute is a 60th Part of a Model in any Order; as fuppofe the Diameter of the Column or Pillar be 18 Inches, then 18 Inches is a Model, and 3 Tenths is a Minute: and hence, if a Column of 18 Inches Diameter be 7 Models, 12 Minutes, it's Height is 10 Foot, 9 Inches, 6 Tenths.

2. In

2. In the Use of the Sector, or which is the same of *Scamozzi's* Lines, which come to the Center or Joint of the Rule, the Word *Lateral* signifies an Extent taken from the Center of the Scale to the Number propos'd, set off upon either Limb of the 30 Scale as Lateral 17 is the Extent from the Joint or Center of the Rule to 17 on the 30 Scale on either Limb; but a parallel Extent is from any Number upon one Limb to the same upon the other: Thus, at whatever Opening the Rule is set, the Extent from 17 on one Limb to 17 on the other is called a Parallel Extent of 17, &c. The Proportion between the Lateral and Parallel Extents upon the Sector, and consequently upon *Scamozzi s* Lines are fully demonstrated, *Euclid Lib.* VI. *Prob.* II. and IV. and upon that Foundation we may build the following general Rule, which is universally true in the Use of all Sectoral Lines, whether equally divided or not, *viz.*

*That every Lateral is to its Parallel, as any other Lateral is to its Parellel*; The Sector or Rule being kept at the same Extent or Opening: Therefore in all Proportions in Common Numbers to be performed by *Scamozzi's* 30 Scale; the Rule is this;

*Make the Lateral Second Term a Parallel in the First, then shall Parallel Third Term be Lateral Fourth required.*

Example.

*Example.* As 18 to 24, so is 12 to what? Make Lateral 24 a Parallel in 18, then shall Parallel 12 be Lateral 16; the Answer, or to be more plain, extend the Compasses from the Center in the Joint of the Rule to the second Term 24, and keeping that Extent in the Compasses, open the *Sector* so as that Extent may reach from 18 on the 30 Scale on one Leg, to 18 on the 30 Scale on the other Leg of the Rule; then keeping the Rule at the same Opening, extend the Compasses from the Third Term 12, on one Limb of the Rule, to 12 on the other Limb, that Extent apply'd from the Center or Joint of the Rule along the 30 Scale, will reach to 16 the Answer required; and this being the Method of working all direct Proportions either in Lines or Numbers, I need not repeat Examples.

The Use of *Scamozzi*'s Lines in several necessary Problems.

PROB. I. *To divide a Given Line less than two Foot into any Number of equal Parts, as suppose a Line of 7 Inches is to be divided into 17 equal Parts.*

TAke 7 Inches in your Compasses, and make it a Parallel in 17 upon the 30 Scale, then shall the Parallel of 10, 12, 15, &c. represent their respective Numbers of Parts upon the Line of 7 Inches proposed to

F      be

be divided. The Ufe of this is, if I would reprefent a Garden of 28 Yards fquare upon a
piece of Paper of 3 Inches fquare, take 3 Inches off the Scale and make it a Parallel in 28,
and keeping the *Scale* at the Opening the
Parallel of 5, 6, 10, *&c.* fhall reprefent the
fame Number of Yards upon the Paper, or to
any Number under 30 ; but if a Number a-
bove 30 be required to be laid down upon a
Line of 7 Inches : As fuppofe 48, take the
half of it, *viz.* 24 ; makes the given Line 7
Inches a Parallel in 24 ; then fhall every Parallel reprefent its double, as the Parallel of
9 reprefents 18 ; and if the Line was to be
continued beyond 48, the Method is the fame,
for the Parallel of 30 would be the Meafure of
60, *&c.* and by this Means without a Multiplicity of Problems any Scale to any Draught
may be made either to meafure, or lay it
down by.

P R O B. II. *To fet the* 30 *Scale at any Angle propofed, or being fet in any Pofition to
know what Angle one* 30 *Line makes with
the other ; and firft to fet it at Right Angles,
or a perfect Square.*

Make Lateral 90 of the Chords, a Parallel
in 15 on the 30 *Scale* ; then does the two 30
*Scales* make a Right Angle.

Or take 10 of the 30 Scale in your Compaffes, and fetting one Foot in 8, open the Rule
till

till the other Foot will fall in 6 on the fame *Scale*, and then the two 30 Lines ftand at Right Angles. *Euclid. Lib. 1. Prob.* 46. Or make Lateral 21 Foot 2 Inches a Parallel in 15, it produces the fame.

To fet the Rule to any Angle propofed ; as fuppofe of 50 Degrees, make Lateral 50 of the Chords a Parallel in 15 on the 30 *Scale*, and it is done ; and to know what Angle the Rule lies open at in any Pofition, make Parallel 15 on the 30 Line a Lateral upon the Chords, and that gives the Angle requir'd.

P R O B. III. *To find a mean Proportional between any two Numbers given, or in plainer Terms, to find an Intermediate Number, whofe Square is equal to the Product of the two given Extreams.*

Set the 30 Scale at Right Angles by *Prob.* 2 and take off the Lateral half Sum of the two given Numbers, and with this Extent, and one Foot in the Lateral half Difference of the two given Numbers, obferve where the other Foot falls upon the 30 Scale on the other Leg of the Rule, for that is the mean Proportional required. Thus the mean Proportional between 4 and 16 is 8, becaufe 4 times 16 is equal to the Square of 8, or as 4 to 8, fo is 8 to 16.

*Note,*

*Note,* One Ufe of the mean Proportional, among many others is, in meafuring Solids, whofe Breadth and Depth are not equal, for in that Cafe the mean Proportional between the Breadth and the Depth is equal, to the Side of the Square, and to be ufed as fuch in all Cafes where the Side of the Square is a Term ufed to find the Content.

**P R O B.  IV.** *Of Superficial Meafure, and firft the Breadth of a Plank given, to find how much in Length makes a Foot.*

Make Lateral 12 a Parallel in the Breadth, and letting the Rule continue fo, take off Parallel 12, and that apply'd laterally gives the Length of a Foot.

*Example.* At 9 Inches broad make Lateral 12 a Parallel in 9, then fhall Parallel 12 be Lateral 16, the Length of a Foot required.

2. *The Breadth of a Board given in Inches, and the Length in Feet, to find the Content in Feet.*

Make Lateral Length in Feet a Parallel in 12, then fhall Parallel Breadth be Lateral Content in Feet.

*Example.* Breadth 8 Inches, Length 18 Feet, make Lateral 18 a Parallel in 12; then will Parallel 8 make Lateral 12, the Content in Feet.

P R O B.

PROB. V. *Of folid Meafure fer Timber; and firft at any Number of Inches fquare, to know how much in Length makes a Foot.*

N. B. *If a Piece of Timber be not fquare, make it a fquare by help of a mean Propcrtional, as in* Prob. 3. *and ufe that as if it was the fide of the fquare.*

Make Lateral Side of the Square a Parallel in 12 (always) and keeping the Rule at that Openirg makes Parallel Side of the Square a Lateral fourth Number. Again, make Lateral 12 a Parallel in that fourth Number, then will Parallel 12 be laterally the Number of Inches to make a Foot.

*Example.* At 9 Inches fquare, how much in Length makes a Foot. Make Lateral 9 a Parallel in 12, then will Parallel 9 be Lateral 6¾. Again, make Lateral 12 a Parallel in 6¾, then will Parallel 12 be Lateral 21¾ Inches to make a Foot of Timber.

2. The Side of the Square in Inches, and the Length in Feet given, to find the Content in Feet and Inches.

Make Lateral Side of a Square a Parallel in 12 (always) then is the Parallel Length a Lateral fourth Number.—Again, make the Lateral fourth Number a Parallel in 12, then is Parallel Side of the Square the Lateral Content in Feet.

*Ex-*

*Ex.* Suppofe a Piece of Timber be 20 Feet long, and 9 Inches fquare, make Lateral 9 a Parallel in 12, then is Parallel 20 (the Length) a Lateral fourth Number, *viz.* 15. Then make Lateral 15 a Parallel in 12, then is Parallel 9 a Lateral 11¼ the Content in Feet required.

## C H A P.  XIII.

*The Ufe of the foregoing Problems in Building, viz. in finding the Lengths and Angles of Rafters, Hips and Collar-Beams in any Square, or Bevelling Roof at any Pitch.*

IN order to the better underftanding the following Directions, it is neceffary to obferve that *True Pitch is when the Rafter is juft three Quarters the Breadth of the Frame or Houfe,* and to find it readily feek the Breadth of the Houfe, upon the 40 Scale, and againft it you will find on the 30 Scale the Length of the Rafter required, three Quarters of which is (nearly) the Height of the Perpendicular from the raifing Piece to the Top of the Gable, which is found on the Rule, by looking for the Length of the Rafter on the 40 Scale, and againft it on the 30 Scale in the Perpendicular of the Gable required.

*Example*

*Example.* Let the Houfe be fuppofed 16
Feet broad, the Rafters according to this
Proportion are to be 12 Feet long, and the
Perpendicular will be 9 Feet.

*Note,* This Proportion is not exactly or ma-
thematically true for making half the Breadth
of the Houfe which is 8, the Bafe of a Right-
angled Triangle; the Length of the Rafter 12,
the Hypotenufe, and the Perpendicular of the
Gable 9, the Perpendicular of the Triangle:
The Sum of the Squares of the Legs is 145,
but the Square of the Hypotenufe 12, is but
144, whereas they fhould be equal, *Euclid.*
*Lib.* 1. *Prob.* 47. but the Difference is fo in-
fenfible that it is thought fufficiently exact in
Practice.

To avoid Multiplicity of Schemes and Problems, as to Square, Bevelling or Hipt Roofs,
I fhall include as much Variety as can be intel-
ligibly done in one; and firft,

**P R O B. I.** *The Confruction of a Roof or*
*Frame that has one End Square and Upright,*
*and the other Bevel and Hipt.* See *Fig.* 1.

Let PMFG reprefent the Frame of a Houfe
whofe Gable End PM is fquare, and the other
End FG is Bevel and Hipt, and let the Square
End PM be 12 Foot the faid MG 23 Feet, the
Bevel End GF 13 Feet 6 Inches, and the Side

F 4                    FP

F P 16 Feet 6 Inches ; take P *v* three-fourths
of the Breadth of the Frame PM in your
Compasses, and one Foot in P draw the Arch
*b*, and with the same Extent and one Foot in
M cross the Arch *b* in *b*, and draw M *b* and
P *b* for the Length of the Rafters 9 Feet,
bisect B M in *a*, and F G in *n*, and draw *n a*
continued to *b*, then *n a* represents the Ridge of
the House, and *a b* 6 Feet 9 Inches (¾ of the
Rafter P *b* which is 9 Feet) is the Perpendicular of the Gable.

Take F *n* or *n* G in your Compasses, and
set it off from *n* to C, and through C draw KF
and G I, which will always cross each other at
Right Angles in C, draw S T perpendicular
to *a n*, and continued both ways to R and V,
and draw *k l* parallel to F G, both through the
Center C.

### To find the Length of the Hips.

Take the Perpendicular *a b*, and set from
C to I and K 6 Feet 9 Inches, and draw K G
and F I ; with the Extent F I, and one Foot
in F make the Arch Y ; and with the Extent
G K and one Foot in G cross the Arch Y in
Y, then draw G Y, it is the longer Hip, and
FY is the shorter.

Continue the Line PM both ways to Q and
Z, and set the Rafters Length M B or PB
from M to Z and from P to Q, as also from S
to R, and from T to V, and draw Z V and
Q R, then is P M *b* the Gable end, P F R Q
one

one Side of the Roof; M Z V G the other
Side of the Roof, and G Y F the Hipt Roof
of the Bevel-end ; and if the Frame P M G F
were supposed to lie horizontally, and the Gable P M *b* set perpendicular upon the Line
P M ; the Side of the Roof M Z V G folded
in the Line M G, and the Side P Q R F fold
up in the Line P F, the Line Z V and Q R
will co-infide and form the Ridge of the Roof,
and the Points R and V will meet perpendicular to the Point C ; and if the Roof of the
Hipt Bevel end F G Y be folded up in the
Line F G, the Point Y will also meet with R
and V, and compleatly cover the Roof, and
the Rafter, Hips, *&c.* may be cut to their re-
fpective Lengths by the Scale by which the
Frame was made, and the Rafters at 12 or 14
Inches afunder, or more or lefs may be reprefented by the Lines *m n, m n, &c.* parallel to
Q P and M Z.

*To find the Back of the Hips to make them*
*anfwer both the Sides and End of the Roof ;*
*and Firft, to find the Back of the longeft*
*Hip G K.*

Lay a Scale from *n* to *l*, it cuts the Diagonal C G, in *x* ; fet one Foot of the Compaffes in *x* and extend the other to the near-
eft Diftance to the Line K G, and fet that
Extent from *x* to *y*, and draw the Lines *l y*
and

and *y n* which will form the Angle of the Back of the longer Hip, and in like manner lay a Scale from *k* to *n*, it cuts F G in *d*, set the Compaſſes from *d* the neareſt Diſtance to F I and ſet that Extent from *d* to *r*, and draw *k r* and *r n*, they make the Angle of the Back of the leſſer Hip.

. But if for private Reaſons the perpendicular Height of the Roof and the Breadth of the Frame be given or determin'd, and by that Means the Roof is above or under true Pitch, increaſe or diminiſh the Perpendicular *a b* according to the given Perpendicular of the Gable, and form the Rafters, Hips, &c. accordingly.

*Example.* Let the Breadth of the Frame be P M as before 12 Foot, but the Perpendicular is limited to 4 Feet 9 Inches, then is *a d* 4 Feet 9 Inches the Perpendicular, and M *d* 7 Feet 7 Inches the Length of the Rafter, equal to which M∠ muſt be made; and from this Foundation proceed in all Reſpects as above directed.

*From what has been ſaid a General Rule may be form'd for Roofing of all Frames, whether above or under true Pitch, the Perpendicular and Breadth being given ; or if true Pitch, the Perpendicular being found by the foregoing Directions, which being known,*

known, *the Method of performing by* Scamozzi's *Lines is as follows* ; And in order
thereto, fet the 30 Scales at Right Angles,
by *Prob.* 2. *Page* 82. then,

### 1. *To find the Length of the Rafters.*

Count half the Breadth of the Houfe on
one Leg, and the Length of the given Perpendicular on the other Leg (both on the 30
Scale (the Parallel Diftance between them meafured Laterally gives the Length of the Rafter
required.

### 2. *To find the Length of the Hip.*

Reckon the Length of the Rafter now found
on one Leg, and half the Breadth of the Houfe
on the other Leg, the parallel Diftance between
them meafured laterally gives the true Length
of the Hip required.

### 3. *To find the Diagonal Line.*

Set off Half the Breadth of the Houfe on
both Legs, and the parallel Diftance between
them meafured laterally gives the Diagonal required.

### 4. *By the Diagonal given to find the Length of the Hips, by a Method different from that in Article* 2.

Set

Set off the Diagonal on one Leg, and the perpendicular Height on the other, and the parallel Diſtance between them meaſured laterally gives the Length of the Hip required.

*Theſe four laſt Articles are ſo plain that I ſuppoſe they need no Example.*

But if the ſquare End be to be hipt, obſerve the following Rules, continue the Perpendicular *a b* to *q*, making *a q* equal to M Z, or Q P the Length of the Rafter, and draw M *q* and P *q*, and ſet the half Breadth *a* M or *a* P from *a* to *t*, then is *a q* the longeſt Rafter of the hipt End, M *q* and P *q* the Hips, *t* the Foot of the King-poſt, perpendicular to which is *i*, *q*, and *i*, will all meet in a Point when folded as before propoſed, and then will the Length of the Ridge be only V *i*, which will coincide with R *i*, as will *i* P with P *q*, and *i m* with M *q*, then will *m o* and *n o* be ſhort Hip Rafters, and P Q *i* and M *z i*, which were of uſe only when the ſquare end was ſuppoſed Perpendicular, are now of no uſe at all.

*By* Scamozzi's *Lines, to find the Hips,* and Height of the King poſt.

Set the 30 Lines at Right angles, then ſet half the Breadth of the Houſe *a* M on one Leg, and the Length of the Rafter *a q* equal to MZ on the other Leg (both laterally from the Center (the parallel Diſtance between theſe two Points meaſured laterally give the Length of

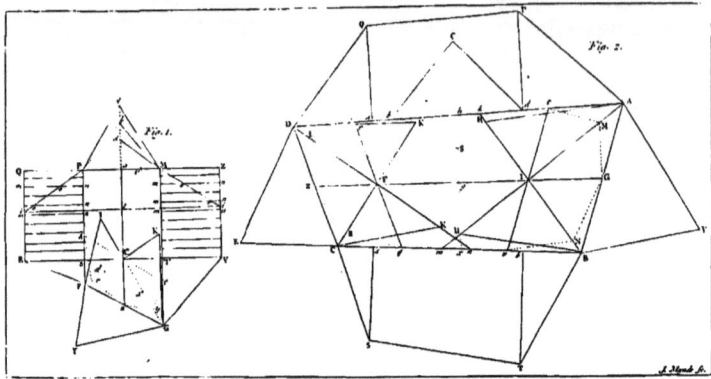

*Fig. 1.*

*Fig. 2.*

of the Hip requir'd. But to find the Height
of the King-poft (having the 30 Scale fet at
Right Angles (take the Lateral Length of the
Rafter *a q* in your Compaffes, and fetting one
in the half Breadth of the Houfe reckon'd laterally from the Center, turn the other Foot,
and obferve where it cuts the 30 Scale on the
other Leg, for that reckon'd from the Center
laterally is the Height of the King-poft requir'd.

*To find the Rafters, Hips and Angles in Frames
that Bevel and are Hipt at both Ends, and
are broader at one End than the other.* See
*Fig* 2.

In thofe fort of Frames the middle Breadth
is the Guide for the Length of the Rafters,
for the Perpendiculars at each End are to be
alike and equal to the middle Perpendicular ;
but in this Cafe the Rafters at the broad End
muft be lefs, and thofe at the narrow End more
than three fourths of the Breadth of their re-
fpective Ends, if they ftand fquare to keep the
Ridge of the Houfe upon a Level or Parallel
to the Horizon ; in order to which let ABCD
reprefent the Frame of the Houfe or Plan it
ftands upon, bevelling at both Ends both at
A B and C D ; where note if the two Sides
and two Ends be given as A D and B C, as alfo A B and D C, yet it is neceffary to have one
of

of the Angles to regulate the reſt. Then biſſect A B in G, and DC in Z, and draw Z G the Line over which the Redge of the Houſe is to be, and upon that Line ſet A G from G to I, and C Z from Z to F; through F draw *r q* parallel to D C, and through I draw *t v* parallel to A B, then through I draw the Diagonals A *m* and B *k*, which will always cut each other at Right Angles; through F alſo draw the Diagonals D *n* and C K, which will alſo cut each other at Right Angles.

Find the middle of the Line Z G as at *g*, in which croſs ZG at Right Angles with the Line *x b*, three fourths of which, *x s* is the Length of the middle Rafter *b* C or C *d*. Now to find the Length of the Hips, and firſt for the Hips ſtanding upon the Angles A and B, A I and B I repreſents the Diſtance of the Foot of the King-poſt from the reſpective Angles A and B, alſo I H is, by Conſtructions equal to the Perpendicular of the Roof or Height of the King-poſt; therefore in the Right Angled Triangle A I H, in which A I is the Baſe I H is the Perpendicular, A H is the Hypotenuſe, from thence called the Hip of the Houſe, therefore A H and B H are the Length of the two Hips for the Angles A and B, and if AV be made equal to A H, and B V equal to B H; theſe two Lines join'd in V the Triangle, A B V exactly repreſents the Roof of the Bevel End A B from Hip to Hip, and by the

the fame Rule the Hips of the other End D C
are found to be D E and C E, and the End of
the Roof to be reprefented by the Triangle
C E D; and if the Hips be meafured in the
Scheme by the fame Scale by which they were
protracted, their Length may be exactly deter-
mined from thence.

For the Angles for the Back of the Hips, I
refer the Reader to the Directions for the Bevel
End of a Houfe in the Explanation of *Fig.* I.

The following Propofitions perform'd by
*Scamozzi*'s Lines.

*To find the Length and Angles of every prin-*
*cipal particular Rafter in Frames broader at*
*one End than the other.*

The Perpendicular being the fame in all
parts of the Houfes as hinted before, open the
30 Scale fquare, and fetting one Foet of the
Compaffes in the Height of the Perpendicular in one Leg of the 30 Scale, and fet the
other Point in the half Breadth of the Frame
on the other Leg, that Extent of the Compaffes
meafured laterally from the Center, gives the
Length of the Rafter required, and to find
the Angles, lay a Rule to the Compafs Points,
whether *Scamozzi*'s, or any other joint Rule,
and fet a Bevel, and it fhews the Angle at the
Raifing piece, or at the Ridge of the Houfe, to
cut the Ends of the Rafters by.

G 2

## To find the Length and Angles of Collar Beams in any Roof.

Take the whole Breadth of the Frame in your Compaſſes meaſured lateraliy on the 30 Scale, and keeping the Compaſſes at that Extent, ſet one Foot in the Length of the Rafter on one Leg, and the other in the Length of the Rafter on the other Leg, that Extent meaſured laterally on the 30 Scale gives the Length of the Raiſing-piece within the Rafters ; then at what Height above the Raiſing-piece you intend the Collar-Beam ſhall be, if you lay a Ruler parallel to the Line between the two Points of the Compaſſes, that Rule ſhall repreſent the Collar-Beam (allowing ſpare Wood for the Tenons) or if it be an Extent of another pair of Compaſſes it is the ſame, meaſured laterally on the 30 Scale gives the Length of the Collar-Beam requir'd.

For the ſides C B T S and D Q P A with the length of their Rafters, &c. See *Prob*. I. *Page* 87, the Manner of performing being much the ſame as in the bevelling Hipt Roof. *Fig*. I.

# F I N I S.